NATO's Mediterranean Initiative

Policy Issues and Dilemmas

F. Stephen Larrabee

Jerrold Green

Ian O. Lesser

Michele Zanini

Prepared for the
Italian Ministry of Defense

National Security Research Division

RAND

The research described in this report was sponsored by the Italian Ministry of Defense.

Library of Congress Cataloging-in-Publication Data

NATO's Mediterranean initiative: policy issues and dilemmas /
F. Stephen Larrabee ... [et al.].
    p.   cm.
"Prepared for the Italian Ministry of Defense by RAND's
National Security Research Division."
"MR-957-IMD."
Includes bibliographical references.
ISBN 0-8330-2605-4
1. National security—Mediterranean Region. 2. Mediterranean
Region—Strategic aspects. 3. North Atlantic Treaty
Organization. I. Larrabee, F. Stephen. II. RAND Corporation.
III. Italy. Ministry of Defense. IV. RAND Corporation.
National Security Research Division.
UA646.55.F88   1998
355 ' 0330182 ' 2—dc21   98-14140
                                                          CIP

RAND is a nonprofit institution that helps improve policy and decisionmaking through research and analysis. RAND's publications do not necessarily reflect the opinions or policies of its research sponsors.

Published 1998 by RAND
1700 Main Street, P.O. Box 2138, Santa Monica, CA 90407-2138
1333 H St., N.W., Washington, D.C. 20005-4707
RAND URL: http://www.rand.org/
To order RAND documents or to obtain additional information,
contact Distribution Services: Telephone: (310) 451-7002;
Fax: (310) 451-6915; Internet: order@rand.org

Since 1989, NATO has focused primarily on Eastern Europe, giving the Mediterranean only limited and sporadic attention. However, in the coming decades, the Mediterranean region is likely to become more important. If NATO and European Union enlargement succeed, East Central Europe will become increasingly stable and integrated into Euro-Atlantic political, economic, and security organizations, decreasing the need for security in Eastern Europe. Security problems will shift to the Alliance's Southern periphery—the Balkans, the Mediterranean, and the Caucasus. Therefore, pressure to address these security problems will likely increase, especially from those members in NATO's Southern Region.

This report examines NATO's Mediterranean Initiative, which was launched at the end of 1994, and how the initiative can be developed as a vehicle for addressing some of the key security problems in the Mediterranean region. The report analyzes the growing connection between Mediterranean and European security; the contribution of other fora, particularly the "Barcelona process" (the European Union's Euro-Mediterranean Partnership, launched in Barcelona in 1995), to enhancing Mediterranean security; the origins and goals of NATO's Mediterranean Initiative; and the perspectives about the initiative of the members of the dialogue. The final chapter suggests ways in which the initiative could be expanded and deepened to enhance transparency of and understanding about NATO's goals and purposes.

This report was written as part of a project on "NATO's Mediterranean Initiative: Policy Issues and Options," sponsored by the Italian Ministry of Defense. Research for the report was carried out in the International Security and Defense Policy Center of RAND's National Security Research Division (NSRD), which conducts research for the U.S. Department of Defense, for other U.S. government agencies, and for other institutions.

The report is based on material from publicly available sources as well as extensive interviews with officials from NATO member states and from the "dialogue countries."[1] It should be of interest to scholars and U.S. and European officials concerned with NATO, European security, and Mediterranean affairs.

Research for this report was completed in October 1997.

---

[1]The term "dialogue countries" refers to the six countries that currently participate in NATO's Mediterranean Initiative: Egypt, Israel, Jordan, Mauritania, Morocco, and Tunisia.

CONTENTS

FIGURE

TABLE

Since 1989, NATO has focused primarily on enlargement in Eastern Europe and on internal adaptation to this enlargement; the Mediterranean has received only sporadic attention. However, in the coming decades, the Mediterranean region is likely to become more important—not just for the Southern members of NATO but for the Alliance as a whole. If NATO and European Union (EU) enlargement succeed, East Central Europe will become increasingly stable and integrated into Euro-Atlantic political, economic, and security institutions. Security problems will likely emerge on the Alliance's Southern periphery—in the Balkans, the Mediterranean, and the Caucasus.

With the end of the Cold War, the locus of risks and challenges is moving south. At the same time, the distinction between European and Mediterranean security is becoming increasingly blurred as a result of the spillover of economic and social problems from the South, such as immigration, terrorism, and drug trafficking, to Europe. Europe is also more exposed to risks from the Middle East.

The growing turmoil in the South will increasingly affect Alliance interests. In addition, the growing involvement of the EU through the Barcelona process (the European Union's Euro-Mediterranean Partnership, launched in Barcelona in 1995) will have an indirect impact on NATO. As the EU becomes more deeply involved in the Mediterranean region, Mediterranean issues will increasingly become part of the European security agenda—and invariably part of NATO's agenda as well. This will make close coordination between the EU and NATO in the Mediterranean more necessary and require the two organizations to work out a more explicit division of labor.

The proliferation of weapons of mass destruction will also thrust problems in the Mediterranean more forcefully onto the NATO agenda. Within the next decade, all the capitals of Southern Europe could be in range of ballistic missiles launched from North Africa and the Middle East. This will create new security dilemmas for these states—and for the Alliance—and could give the security dialogue with these states quite a different character.

Finally, the Greek-Turkish dispute over the Aegean and Cyprus is likely to remain a source of concern and keep the Alliance's attention focused on the Mediterranean. As long as these issues remain unresolved, there is always a danger that some incident could lead to a new confrontation, as almost happened as a result of the flare-up over the Aegean islet of Imia/Kardak in February 1996.

Thus, for many reasons, NATO will be forced to pay greater attention to challenges in the Mediterranean. *The real issue, therefore, is not whether NATO should have a Mediterranean policy but what the nature and content of that policy should be and how it can be most effectively implemented.*

NATO'S MEDITERRANEAN INITIATIVE AT THE CROSSROADS

NATO's Mediterranean Initiative, launched at the end of 1994, signaled the Alliance's recognition of the growing importance of the security challenges in the Mediterranean. However, progress in developing the initiative has been slow.

Several factors have influenced the low profile adopted by NATO toward the Mediterranean. First, the Mediterranean Initiative lacks strong Alliance-wide support. Most members of NATO are willing to support the initiative as long as it is limited to "dialogue" and does not require any increased expenditure of resources. However, except among the Southern members, there is little strong enthusiasm for the initiative within the Alliance.

Second, the goals of the initiative remain ill-defined. It is not clear whether the main purpose of it is simply to conduct a dialogue with the countries of the Southern Mediterranean or whether the initiative is also part of the broader effort to establish defense cooperation with these countries. This lack of clarity, in large part, reflects the limited consensus within NATO about what the initiative is really supposed to do.

Third, NATO has had other, more pressing priorities—particularly enlargement, internal adaptation to enlargement, and development of a viable partnership with Russia. Many countries do not want to see attention and scarce resources diverted from enlargement and the Partnership for Peace (PfP).

Fourth, NATO suffers from a serious "image" problem in the "dialogue countries,"[1] especially among the broader public. The publics in many dialogue countries view NATO as a Cold War institution that is now searching for a new enemy. As a result, the governments in many dialogue countries are wary of cooperating too closely with NATO, especially in the security and defense area, fearing that this will spark a hostile reaction among key segments of their publics. In addition, many are not clear how NATO can help them resolve their security problems, most of which are internal and economic and social.

Fifth, the initiative is not tied to NATO's broader security and defense agenda in the Mediterranean. This agenda involves important security issues such as counterproliferation, counterterrorism, and peacekeeping and humanitarian assistance. However, the relationship between this broader security and defense agenda and the Mediterranean Initiative is unclear. Moreover, many of the dialogue countries prefer to concentrate on "soft" security issues (e.g., those involving migration problems and cultural security) and economic issues rather than on "hard" security issues (e.g., proliferation of weapons of mass destruction and defense cooperation). This imposes serious limits on the dialogue between NATO and each of these countries.

---

[1]The term "dialogue countries" refers to the six countries that currently participate in NATO's Mediterranean Initiative: Egypt, Israel, Jordan, Mauritania, Morocco, and Tunisia.

Sixth, the deterioration of the Arab-Israeli peace process has diminished the willingness of many Arab countries to engage in a dialogue with Israel. This has inhibited progress in NATO's Mediterranean Initiative and made it more difficult to engage many of the North African and Middle Eastern countries in discussions in which Israel is included.

Finally, NATO's Mediterranean Initiative is not tied to the broader U.S. strategic agenda in the Mediterranean and the South Mediterranean more generally. In addition, the United States remains concerned that the initiative will interfere with the Middle East peace process and divert attention from Eastern European enlargement. Hence, Washington has expressed only perfunctory interest in the initiative. However, without strong U.S. support, the initiative is unlikely to amount to much.

As a result of all these factors, NATO's Mediterranean Initiative has not really gotten off the ground. It remains largely an afterthought rather than a serious Alliance initiative with strong political support and momentum. If it is to succeed, the initiative needs to be reinvigorated with stronger political backing, particularly by the United States. It also needs to be more closely harmonized with other initiatives in the region, particularly the Barcelona process. Otherwise, it is likely to become yet another in a long line of failed Western initiatives in the Mediterranean.

HARD SECURITY VERSUS SOFT SECURITY

In addition, the Alliance is confronted with a major contradiction in its dealings with the countries of the South Mediterranean littoral. NATO's comparative advantage is in the area of hard security. The dialogue countries, however, are primarily interested in "soft security." This raises an important dilemma regarding how to structure the dialogue. What should the main priority in the dialogue be: hard security or soft security?

Given NATO's poor image in many of the dialogue countries and the sensitivity about hard security issues in these states, it may be better in the early stages of discussions with many of the dialogue countries for NATO to concentrate on soft security and building confidence rather than on moving directly to defense and military cooperation.

The main reason for initially concentrating on soft security is political/psychological. As noted, many of the dialogue countries have strong reservations about NATO. There is thus a need to develop a "bottom up" approach—to first develop trust and confidence. This can lay the groundwork for the later development of concrete military cooperation.

The best means for helping to develop such trust and confidence would be to expand considerably the participation of the dialogue countries in seminars on security issues of mutual interest and to invite representatives of dialogue countries to attend NATO-sponsored events, including military exercises. This would help to develop greater *transparency* and could lead to the development of a more positive image of NATO in the dialogue countries over time.

This, in turn, could lay the groundwork for the development of military cooperation at a later stage. Such military cooperation, however, should be carried out on a *case-by-case* basis. In effect, the same principle of "self-differentiation," which guides PfP, should be used as a guideline for the Mediterranean Initiative. This would allow for recognition of the differences among the various dialogue countries and allow military cooperation to develop at its own natural pace. Some countries, such as Egypt and Israel, may be willing to develop some low-level forms of military cooperation, particularly in the areas of civil emergency planning, peacekeeping, and peace support activities, while others may not feel comfortable with such cooperation for quite a while. And some may never want it.

NATO, however, should not be the *demandeur*. Rather, the Alliance should allow cooperation to develop at a pace with which each dialogue country feels comfortable. At the same time, NATO should be careful to ensure that military cooperation with Israel does not get too far ahead of military cooperation with other Arab states, especially Egypt and Jordan. This could undermine the effectiveness and objectives of the Mediterranean Initiative.

ARMS CONTROL AND CONFIDENCE-BUILDING MEASURES

Some low-level arms control and confidence-building measures (CBMs) may also contribute to enhancing transparency and building trust. However, the Alliance should recognize that the "security culture" in the dialogue countries differs markedly from that in Europe and should recognize that the experience of the Organization for Security and Cooperation (OSCE), which has contributed to enhancing security and cooperation in Europe, cannot automatically be transferred to the Middle East and North Africa, where the political and psychological environment is quite different. Many Arab states feel that CSCE/OSCE-type CBMs are at present premature and can be introduced only after the conclusion of a Middle East peace settlement. Otherwise they fear such measures will perpetuate the status quo that the Arabs are trying to change. Moreover, many of the security problems in the Mediterranean are of an internal nature. They are not amenable to resolution through classical arms control and confidence-building measures.

This does not mean that confidence-building measures have no utility or should not be tried, but rather that different types of confidence-building measures are needed—at least in the initial stages—from the classical Western measures developed in the CSCE/OSCE context. Rather than trying to introduce arms limitations zones and other similar military CBMs, it might be better to concentrate initially on measures designed to increase transparency and defuse threat perceptions—measures such as security seminars, educational visits, and inviting dialogue country representatives to observe military exercises. This could lay the groundwork for more robust military CBMs—including participation in military exercises—later on.

PfP AND THE MEDITERRANEAN

Some Alliance members have suggested that PfP should be extended to the Mediterranean. They point to the success of PfP in Eastern Europe and argue that extending it to the Mediterranean could help to reduce threat perceptions and foster closer cooperation in the region. On the surface, this idea has considerable attraction. PfP has worked remarkably well in Eastern Europe and has been more successful than most observers and NATO officials initially anticipated. One may legitimately ask, Why not extend it to the Mediterranean?

However, in developing cooperation with the countries of the Southern Mediterranean, the Alliance needs to be sensitive to the very different political and cultural environment that exists in North Africa and the Middle East in comparison to that in Eastern Europe. What works in Eastern Europe may not necessarily work, or be desirable, in the Middle East and North Africa.

This does not mean that PfP has no relevance for the Mediterranean. On the contrary, *some aspects* may be relevant. However, PfP cannot be transferred lock, stock, and barrel to the Mediterranean. In general, the political diversity of the region will require separate and specific solutions appropriate to the region.

THE EFFECT OF THE MIDDLE EAST PEACE PROCESS

NATO's ability to move forward with its Mediterranean Initiative will also be affected by external factors, particularly the Middle East peace process. While formally there is no direct link between NATO's Mediterranean Initiative and the peace process in the Middle East, developments within the Middle East peace process affect the willingness of dialogue countries to cooperate with Israel and even participate in multilateral meetings where Israel is present. Thus, the two issues cannot be entirely separated, even if no formal connection between the two exists.

NATO needs to recognize this broader linkage, since it affects the degree to which the dialogue with countries in the Middle East and North Africa can be deepened and broadened. This does not mean that NATO should not try to develop a more robust dialogue with the countries of the Middle East and North Africa. But it should recognize the degree to which outside developments—above all, progress in the Middle East peace process—affect the prospects for the initiative's success. If the peace process can be reactivated, a broader and more meaningful dialogue with the countries of the Middle East and North Africa will be more likely. But if the peace process remains stalled, developing such a dialogue will be difficult.

MEMBERSHIP IN THE DIALOGUE

At present, the dialogue is limited to the six countries mentioned above: Egypt, Israel, Jordan, Mauritania, Morocco, and Tunisia. Some NATO members have suggested that the dialogue should be expanded.

In principle, there is no fundamental obstacle or objection to expanding the dialogue. The NATO Council's approach has been that the Mediterranean Initiative is open and can be progressively expanded on a case-by-case basis. For instance, Jordan joined at the end of 1995.

However, the experience of the EU's Euro-Mediterranean Partnership suggests that there are merits in keeping the dialogue small and focused. If the dialogue becomes too large, it will be difficult to achieve a consensus among participants, especially on sensitive defense and security issues. Moreover, the larger the membership, the more likely it is that extraneous issues, such as the Arab-Israeli dispute, will be injected into the dialogue, slowing progress. This is particularly true if Syria and Lebanon were to be invited to join the dialogue. Thus, keeping the group relatively small has advantages.

BILATERALISM VERSUS MULTILATERALISM

To date the dialogue has been largely bilateral, except in the field of information. However, many of the dialogue countries do not seem ready for a multilateral dialogue, especially as long as relations with Israel are strained. Moving too quickly to make the dialogue multilateral, therefore, could be disruptive and may reinforce existing tensions within the group. Thus, at this time, it would seem advisable to continue most activities on a bilateral basis. However, some activities could be conducted on a multilateral basis. Multilateral seminars devoted to security issues, for instance, can contribute to creating a broader security community in the Mediterranean and to fostering greater regional cooperation in a variety of areas.

COORDINATION WITH OTHER INITIATIVES

The wide variety of organizations and institutions dealing with various aspects of the Mediterranean poses a special challenge for NATO. In thinking about the future of its Mediterranean Initiative, NATO needs to consider how its initiative fits into the broader pattern of other ongoing efforts aimed at enhancing cooperation and security in the Mediterranean (as discussed in Chapter Two). Given that the main security problems in the region are internal and have their roots in economic, social, and political factors, it makes sense for the EU to take the lead in discussions about the Mediterranean. The EU is much better equipped than NATO to deal with the problems in the region. NATO should not try to duplicate the EU's efforts, but rather complement and reinforce them by exploiting its comparative advantage—which is in the areas of hard security and the military field.

At the same time, the large number of institutions and organizations dealing with Mediterranean security issues increases the importance of broadly coordinating NATO's efforts with those of other institutions, especially the EU and WEU, in order to avoid duplication. The political and military dialogues conducted by NATO and the WEU, for instance, have largely the same goals and include nearly the same countries. Security issues, especially CBMs, are also part of the Euro-Mediterranean dialogue within the Barcelona process.

Initially there may be bureaucratic resistance within the EU—as well as within NATO—to coordinating the two dialogues. But some loose coordination is essential if duplication and confusion are to be avoided. Otherwise there is a danger that the two dialogues could work at cross purposes and undercut, rather than complement, one another.

This coordination need not require the establishment of an elaborate institutional mechanism. It could be done through an informal periodic exchange of views—perhaps two or three times a year—between the Secretary General of NATO and the head of the EU Commission and/or EU commissioners in charge of Mediterranean issues.

AREAS FOR EXPANDED COOPERATION

In addition, a number of steps could be taken to give the Mediterranean Initiative new momentum and relevance in specific areas. Most of these steps are relatively modest and would not require a major expenditure of additional resources. Four areas in particular should be given top priority:

- Public information and outreach
- Educational courses and visits

- Civil emergency planning (CEP)

- Crisis management, peacekeeping, and peace support activities.

In particular, NATO's public information and outreach effort needs to be expanded. This is critical if NATO is to change perceptions in the dialogue countries and create better understanding of the Alliance's goals and purposes. It is also an essential building block for broader cooperation in other areas over the long run. At the same time, NATO should do more to involve dialogue countries in CEP, educational courses and visits, and peacekeeping and peace support activities. These are areas where NATO has a comparative advantage over other institutions and where many of the dialogue countries have expressed an interest in greater cooperation. However, if the Mediterranean Initiative is to succeed, NATO will have to devote greater financial resources to it. Even the modest initiatives suggested in this study will require some increase in funding, especially in the area of information dissemination.

The Mediterranean Cooperation Group (MCG), established at the Madrid summit in July 1997, should be the main body for developing and coordinating NATO's Mediterranean Initiative, except in the field of public information, which should be coordinated by the officer in charge of Mediterranean affairs in the NATO Office of Information and Press. The MCG should be tasked to undertake a major review of NATO's Mediterranean Initiative and to recommend areas where cooperation and dialogue can be further developed. In addition, the MCG should meet periodically—perhaps twice a year—with the dialogue countries to review progress and develop new initiatives.

TOWARD A BROADER SOUTHERN STRATEGY

If the Mediterranean Initiative is going to have a significant impact over the long run, it must be linked with a broader Alliance strategy toward the South. This broader strategy will require linking a number of diverse elements into a coherent whole:

- command reform

- PfP (for candidate and noncandidate members)

- policy toward the Balkans

- dialogue with the countries along the Southern Mediterranean littoral

- enhancing cooperation and stability in the Aegean

- counterproliferation.

At present these various elements are running on separate tracks. They need to be integrated into a broader, more coherent strategy designed to enable NATO to better meet the security challenges it will face in the coming decades, many of which are likely to be on the Alliance's Southern periphery.

In particular, the role of AFSOUTH needs to be upgraded as part of the overall effort to modernize and streamline NATO's command structure. With the end of the Cold War, the responsibilities and importance of AFSOUTH have increased significantly. Yet there has been little corresponding shift in resources to enable the AFSOUTH commander to carry out these expanded responsibilities or to closely monitor and plan for contingencies in his Area Of Strategic Interest (AOSI).

This imbalance needs to be redressed as NATO carries out the process of internal adaptation. More resources need to be shifted to AFSOUTH to enable the AFSOUTH commander to carry out his expanded responsibilities and to monitor and plan for contingencies in his AOSI. In addition, the command reform under way needs to be designed to meet the new challenges NATO is likely to face in the future, many of which are likely to be in the South.

At the same time, the Alliance needs to ensure that its political and military strategy in the Mediterranean are in harmony. Command reform, especially if it is accompanied by an effort to create greater power projection capabilities, could create new anxieties and fears among the dialogue countries and inhibit efforts to intensify cooperation with them. Thus, it is important that any changes in NATO's military strategy and command structure be carefully explained in advance to dialogue countries to reduce the chances of misperception or misunderstanding. This underscores the need to give greater priority to an expanded public information and outreach program.

Embedding the Mediterranean Initiative in a broader Southern strategy would also help to ensure strong U.S. support. To date, the United States has not exhibited a strong interest in the Mediterranean Initiative, in part because it does not clearly perceive the link between the initiative and the "big" strategic issues in Southern Europe. The more NATO's initiative can be linked to the broader U.S. agenda in the Mediterranean, the more likely it is to obtain backing in Washington. Such support will be crucial if the Mediterranean Initiative is to gain political weight and momentum.

ACKNOWLEDGMENTS

The authors would like to express their appreciation to Admiral Giampaolo Di Paola of the Italian Ministry of Defense; General Giuseppe Cucchi, former Director of the Centro Militare di Studi Strategici (CeMiSS) in Rome; Giovanni Jannuzzi, Italian Ambassador to NATO; and Alvaro de Vasconcelos, Director of the Institute of Strategic and International Studies (IEEI) in Lisbon, for their help in the preparation of this report. Special thanks go to Nicola de Santis, Officer for Southern and Eastern Mediterranean Countries of the NATO Office of Information and Press, in Brussels, whose advice, initiative, and assistance were essential to the conduct of the study.

Chapter One

THE CHANGING MEDITERRANEAN
SECURITY AGENDA

During the Cold War, the attention of Western policymakers was focused primarily on the Central Front. The Mediterranean was regarded as of secondary importance. However, in recent years—and especially since the end of the Cold War—the Mediterranean has assumed new importance as a focal point of Western concern. The renaissance of the Mediterranean in security terms is based on its growing role in the strategic calculus of Europe, the United States, and the Middle East. The prosperity and security of key states are increasingly affected by events around the Mediterranean, and this area's capacity for producing crises as well as slow-moving challenges—with potentially far-reaching consequences—has begun to compel the attention of analysts and policymakers.

Many developments could still derail this trend toward greater interest in the Mediterranean, including the rise of new tensions with Russia and insecurity in Eastern Europe, not to mention adverse developments further afield.[1] For the moment, however, Mediterranean issues have begun to occupy a more prominent place in security debates, and are imposing new intellectual and policy challenges on both sides of the Atlantic, and on both shores of the

---

[1]In the event of renewed friction with Russia, it is possible that this strategic competition might be played out through policies and proxies on the periphery of Europe, including the Mediterranean. For some Turkish observers, the Russian sale of military equipment to Cyprus is an early indication in this regard.

Mediterranean.[2] Hence new approaches to security and cooperation in the Mediterranean are likely to become an important part of reform and adaptation within the Alliance.

RETHINKING MEDITERRANEAN SECURITY

Some Western observers have been openly skeptical of the notion of Mediterranean security, arguing that the Mediterranean is too diverse a region in security terms, with a wide range of serious but highly differentiated subregional problems. What, if anything, do the Western Sahara, the Levant, the Aegean, and the Balkans have in common that might suggest a useful "Mediterranean" approach? Moreover, the traditional intellectual (and bureaucratic) divide between European and Middle Eastern affairs makes the development of a Mediterranean approach difficult, especially from the U.S. perspective.

There are, however, compelling reasons to take a broader view. First, the existence of distinctive subregional issues does not eliminate the importance of broader, regional—indeed transregional—approaches to security problems, many aspects of which cannot be adequately understood by viewing issues and crises in isolation. Western policymakers have no difficulty in accepting that Baltic, Balkan, and Central European issues belong within a European security framework, or that events in North Africa and the Persian Gulf contribute to a wider Middle Eastern security environment. Why not employ a Mediterranean lens when the issues and policy responses warrant it? Second, and without losing sight of the specifics, it is clear that many of the security challenges around the Mediterranean basin spring from similar trends—from unresolved questions of political legitimacy, relentless urbanization, and slow growth to resurgent nationalism, religious radicalism, and the search for regional "weight."

---

[2]For a recent discussion, see "Western Approaches to the Mediterranean," (several articles), *Mediterranean Politics*, Vol. 1, No. 2, Autumn 1996. See also John W. Holmes, ed., *Maelstrom: The United States, Southern Europe and the Challenges of the Mediterranean* (Cambridge: World Peace Foundation, 1995) and Ian O. Lesser, *Mediterranean Security: New Perspectives and Implications for U.S. Policy* (Santa Monica, Calif.: RAND, R-4178-AF, 1992).

Third, and above all, the growing interdependence of traditionally separate security environments, as a result of the expanded reach of modern military and information systems, is producing a significant gray area of problems that are neither strictly European nor Middle Eastern. The Mediterranean is at the center of this phenomenon, and Mediterranean security is likely to be an increasingly useful organizing principle for governments and institutions seeking to improve the overall security climate. As NATO itself becomes more actively engaged in addressing functional security issues (e.g., proliferation, terrorism, crisis management) that cut across traditional geographic divides, the importance of the Mediterranean will be reinforced as a natural sphere for action on Europe's doorstep.

THE NEW MEDITERRANEAN SECURITY AGENDA

Most discussion of the security environment in the region rightly encompasses both "hard" (e.g., proliferation of weapons of mass destruction (WMD), drug trafficking, and terrorism) as well as "soft" (e.g., political, economic, and social) issues. Indeed, the expansion of the security agenda beyond narrowly defined defense questions has been a leading feature of the post–Cold War scene everywhere, and the Mediterranean is an example of this trend. It has been argued, with some merit, that the definition of some soft issues, especially migration, as security challenges encourages an overheated treatment by publics and policymakers on both sides of the Mediterranean. Rightly or wrongly, however, migration has emerged as a security issue in European perceptions. At the same time, opinion elites in North Africa, as well as Turkey, are coming to regard the treatment of their compatriots in Europe as part of the foreign and security policy agenda in the broadest sense.

Many analysts of Mediterranean affairs point to the growing gap between a "rich" North and a "poor" and increasingly populous South. The population around the Southern and Eastern Mediterranean is likely to reach 350 million by the end of the century. By contrast, the total population of the current members of the European Union (EU) will probably not exceed 300 million in the same period. Over the last decade, the countries of the Maghreb, including Egypt, have experienced population growth on the order of 40 percent. Although there is now some indication that these tremendous rates of growth have slowed, the trend has clearly been toward a growing imbalance of population between the North and South—indeed a virtual reversal of the demographic weight of Europe in relation to North Africa and the Levant. From a social and political viewpoint, it is perhaps more telling that the number of people under 15 years of age near the Southern and Eastern shores of the Mediterranean will reach some 30 percent by 2025.[3] Demographic pressures of this magnitude are producing relentless urbanization, social and economic strains, and a steady stream of migrants seeking jobs and social services (a process which starts well to the south of the Maghreb and affects societies on *both* sides of the Mediterranean). Needless to say, these pressures, together with more visible divides between the "haves" and the "have-nots," are also threatening the political stability of states around the Southern Mediterranean.

Energy issues have more commonly appeared on "Northern" agendas as a security concern, but with the growth of new lines of communication for energy around the Mediterranean basin, the interest in energy security is now more broadly shared. The discussion of energy as a security issue is also losing some of its traditional North-South character, as South-South energy links expand (e.g., the trans-Maghreb gas pipeline, Iranian-Turkish links, and alternative routes for bringing Caspian oil to Western markets). The development of new energy ties may be seen in some quarters as a source of additional vulnerability to political turmoil beyond national and regional borders. Yet new links are just as likely to have a stabilizing effect because of diversification and economic interdependence along both South-South and North-South axes. In this context, it is not surprising that the Euro-Mediterranean Partnership (also known as the "Barcelona process") has made cooperation on energy issues a focus for early attention. Dialogue on the security dimensions of energy trade and use could emerge as a promising area for NATO in its own Mediterranean Initiative.

The softest part of the Mediterranean security agenda, but one arguably increasing in significance, concerns what can be termed

---

[3]This discussion is based on World Bank and United Nations estimates. See *United Nations World Economic and Social Survey 1995;* and Eduard Bos et al., *World Population Projections 1994–95* (Baltimore, Md.: Johns Hopkins University Press, 1994).

"security of identity."[4] Security of identity, or cultural security, is a highly charged issue in many Mediterranean societies and has been prominent in the thinking of secular as well as religious observers in North Africa and the Middle East. It is also implicit in speculation about "civilizational" clashes, with the Mediterranean as a leading fault line between Islam and the West. The widespread availability of Western television and other media has heightened awareness of the identity issue. Migration from South to North has introduced another sort of concern about the meaning of immigration for the cultural security of recipient states. This anxiety has contributed to the politicized debate over immigration policy in Mediterranean Europe, reinforcing the economic and security aspects of the issue. Without judging the validity of cultural anxieties on both sides of the Mediterranean, it is clear that perceptions about security of identity can have a marked effect on the prospects for Mediterranean dialogue and cooperation on other fronts. This issue transcends the well-known problem of the divide between public and elite opinion in North Africa and elsewhere in the "South." Ambivalence toward, or even outright distrust of, Western institutions is observable even among certain Southern Mediterranean elites, a reality reinforced by perceived challenges to identity. Any attempt to deepen NATO's engagement and dialogue across the Mediterranean will need to address issues of identity as part of a broader public information strategy.

"Hard" security problems in the military and defense realm are similarly diverse. These range from political violence and terrorism, to the proliferation of WMD and longer-range weapons delivery systems. Less fashionable, but still central to the Mediterranean security environment, are the sophisticated, large-scale conventional arsenals and the challenges these pose to the territorial status quo. Despite the looming exposure of Europe to longer-range weapons deployed around the Mediterranean periphery, the direct military risks remain largely South-South rather than South-North, especially in the Western Mediterranean. There are varying but identifiable risks of military confrontation between Morocco and Algeria, Libya

---

[4]See Fernanda Faria and Alvaro de Vasconcelos, "Security in North Africa: Ambiguity and Reality," *Chaillot Paper,* No. 25 (Western European Union (WEU) Institute for Security Studies, September 1996), p. 5.

and Tunisia, Libya and Egypt, and Egypt and Sudan. Direct confrontations across the Mediterranean are more difficult to imagine under current political conditions, with the possible but remote exception of a Spanish-Moroccan crisis over the enclaves of Ceuta and Melilla.

In the Eastern Mediterranean, the potential for large-scale armed conflict is more prominent in the current strategic environment. The Arab-Israeli dispute continues to have an important military dimension, both conventional and unconventional. The risk of conflict between Greece and Turkey remains high, and the circumstances have acquired additional gravity with the recent increases in tension in the Aegean and about Cyprus. If the Balkans and the Black Sea region are included in the equation, the potential for armed conflict in and around the Mediterranean is far from theoretical.

Looking across the Mediterranean security agenda, one point that emerges very strongly is the extent to which individual crises (e.g., in Bosnia, Algeria, the Aegean, Arab-Israeli relations) can influence security perceptions across the region. In addition, deterioration in the climate surrounding political, economic, or even cultural issues could produce an environment in which more-direct security risks increase, crises become more difficult to manage, and initiatives aimed at Mediterranean dialogue become more controversial. Many aspects of the Mediterranean security equation are unrelated to the Arab-Israeli relationship. Nonetheless, it will be impossible to ignore the links between the health of the Middle East peace process and the prospects for deepening security cooperation along Mediterranean North-South lines. In terms of giving NATO's Mediterranean Initiative a true multilateral (as opposed to bilateral or "multi-bilateral") character, the outlook is likely to be heavily influenced by the overall tenor of Arab-Israeli relations.

THE INTERNAL DIMENSION

For many states around the Mediterranean, security continues to be, above all, a matter of *internal* security, and many foreign and security policy questions derive importance from their ability to affect the stability of existing regimes. Along the Southern and Eastern shores of the Mediterranean, political futures remain unresolved, with many regimes facing significant challenges to their legitimacy. The near–

civil war in Algeria provides the most dramatic example of internal insecurity and violent Islamist opposition to the political order. Whether or not the Algerian regime succeeds in containing the Islamic insurgency, the Algerian crisis is likely to have a profound effect on the security of North Africa as a whole, and the overall perception of risk from the South in Mediterranean Europe.[5] Regardless of the outcome, Algerian society will be profoundly changed, and Algeria's neighbors on both sides of the Mediterranean may face spillovers of terrorism and political extremism for some time to come. The Algerian crisis has thrown the question of political change and the role of Islam along the Southern shores of the Mediterranean into sharp relief. In Morocco, Tunisia, Egypt, and even Libya, security perceptions will be driven by the need to preserve political legitimacy and hold violent (or potentially violent) opposition movements at bay.

The problem of political legitimacy and internal stability will be closely tied to demographic and economic trends across the region. The dilemmas posed by expanding and younger populations coupled with slow economic growth have been widely discussed. From Morocco to Turkey, attempts at economic reform and the emergence of a more dynamic private sector are widening the gap between the "haves" and the "have-nots," with potentially destabilizing consequences. Reforms aimed at promoting longer-term prosperity and encouraging foreign investment may well reinforce stability over the longer term, but the shorter-term political risks are substantial, especially where dissatisfaction with the existing political order is already widespread. Rising expectations will be difficult to meet and could prove to be a powerful source of political change in countries where the established political class proves incapable of promoting a better distribution of wealth and opportunity. In the Eastern Mediterranean, the rise of Turkey's Islamist Refah Party provides a striking example of the political consequences of uneven increases in prosperity and the failure to address pressing social problems.

---

[5]See Graham E. Fuller, *Algeria: The Next Fundamentalist State?* (Santa Monica, Calif.: RAND, MR-733-A, 1996); Andrew J. Pierre and William B. Quandt, *The Algerian Crisis: Policy Options for the West* (Washington, D.C.: Carnegie Endowment, 1996); and Simon Serfaty, "Algeria Unhinged," *Survival*, Vol. 38, No. 4, Winter 1996–97, pp. 137–153.

These political and economic stresses have been compounded by the relentless urbanization affecting virtually all Mediterranean societies. The Southern and Eastern shores of the Mediterranean are among the most highly urbanized areas in the world, with cities such as Istanbul and Cairo experiencing extraordinary rates of growth over the last few decades. Urbanization has shaken traditional patterns of behavior and placed enormous new demands on already hard-pressed governments. The inability of governments to meet the needs of urban populations has led to an increasing tendency of urban citizens to organize their lives without reference to the state and has provided an opening to Islamist movements with effective municipal organizations. In Algeria, Egypt, and Turkey, urban dissatisfaction and the Islamic movements' ability to provide services unavailable from the state have been significant sources of power for Islamic activists. In security terms, continued urbanization suggests an environment in which cities will be the focal point for instability and opposition and centers of political rivalry, both violent and nonviolent. If security concerns across much of the Mediterranean will be about internal security, cities will be the focus of insecurity within societies where insecurity is already pervasive.

Much of the foregoing discussion has concentrated on the problems of the South. But societies on both sides of the Mediterranean share in a growing perception of declining "personal security." In places as diverse as Algeria, Bosnia, and Southeastern Anatolia, the threats to personal security are direct and obvious. In Israel, the result of the last national elections can be regarded less as a referendum on the peace process than on the question of personal security in the wake of terrorist actions. In Southern Europe, and Europe as a whole, the concern about spillovers of political violence from crises across the Mediterranean compels the attention of political leaderships and public opinion because terrorist risks strike at personal security as well as at the security of the state. The terrorist bombings in Paris in 1996 provide a dramatic example. In France and elsewhere, right-wing movements have used the personal security issue (crime, terrorism, drug trafficking), in addition to economic and "identity" arguments in support of their views on immigration policy.

A detailed analysis of the implications of the information revolution for security in the region is beyond the scope of this discussion, but three aspects are worth mentioning. First, the growing ease of telecommunications is likely to bolster the power and flexibility of opposition movements, both violent and nonviolent, within Mediterranean states and in "exile," with implications for the stability of many regimes in North Africa and the Levant. Second, it will facilitate the growth of political networks, including terrorist and criminal networks around the Mediterranean and beyond.[6] As a consequence, the potential for spillovers of political violence (e.g., Algerian Armed Islamic Group (GIA) terrorism in France, Kurdish Workers' Party (PKK) fundraising, and violence in Germany) will increase and the decentralized and freelance behavior of "networked" groups will be difficult to monitor and counter. Finally, the widespread availability of European media around the Mediterranean has already had a marked effect on "Southern" images of the rich societies to the North. Islamists as well as many Arab secularists have seized on this phenomenon as a threat to their security of identity, as noted earlier. Each of these information-related trends will reinforce the interdependence of societies around the Mediterranean and elsewhere in security terms and increase the need for dialogue and practical cooperation.

The pressures for political and economic change in Mediterranean societies will be accommodated in different ways and with different degrees of success. Given the experience of Algeria and the lower-level crises ongoing elsewhere, from the Western Sahara to the Caucasus, however, it is reasonable to expect that the future Mediterranean security environment will be characterized by multiple instances of turmoil within societies, with the attendant risk of spillovers. Whether demographic pressures and internal instability

---

[6]The rise of Mediterranean networks will, of course, have a benign aspect as well. There is a striking parallel between (1) the notion of a Mediterranean region in which like-minded groups, regardless of location, have more in common and more communication with each other than with dissimilar groups within their own societies and (2) Fernand Braudel's description of the traditional Mediterranean world. In his analysis, societies around the Mediterranean shore shared interests and behavior—and had a greater degree of contact—than such societies had with communities in the Mediterranean hinterland. Climate, ease of communications, and commercial interests were more significant than sheer proximity. Differences in altitude and the difficulty of overland travel made trans-Mediterranean communication easier and more attractive than communication with the nearby hinterland. In this environment, the Mediterranean served as a bridge rather than a barrier, and maritime networks flourished. See Fernand Braudel, *The Mediterranean and the Mediterranean World in the Age of Philip II* (New York: Harper and Row, 1966).

lead to the pattern of chaotic violence and failed states characterized by Robert Kaplan as "the coming anarchy," the Mediterranean basin certainly includes a number of societies where outcomes along these lines are possible.[7] In Algeria, "chaos" is already a reality. Looking further afield, two Mediterranean states, Turkey and Egypt, are also experiencing the security consequences of chaos on their borders, the former in the context of turmoil in the Caucasus and Northern Iraq, the latter in relation to conflict and refugee pressures in Sudan.

THE REGIONAL DIMENSION

The combination of internal political change and the continuing effects of the loss of Cold War moorings will have significant consequences for the strategic environment around the Mediterranean, and within key subregions. Some broad trends are worth noting in this context. First, unstable societies and changing political orientations will complicate crisis prevention and management. As an example, radical ideology and humanitarian strains in Sudan increase the likelihood of conflict with Egypt over such substantive issues as water. Similarly, the growing prominence of Islamic politics in Turkey tends to reinforce existing perceptions of a civilizational cleavage between Islam and Orthodoxy, further complicating relations with Greece and Russia, and fueling nationalist instincts on all sides. The advent of new Islamic regimes in Algeria, or perhaps a post-Qadhafi Libya, would give an ideological edge to potential frictions with neighboring states over territorial and other issues.

Second, it has become fashionable to see political Islam as a key driver of internal and external challenges around the Southern and Eastern Mediterranean. Islam is indeed likely to be a continuing and significant force in the political evolution of many states in the region, and a factor in foreign and security policy orientations.[8] But it would be unwise to dismiss the power of nationalism as a key moti-

---

[7]See Robert D. Kaplan, "The Coming Anarchy," *The Atlantic Monthly*, February 1994, pp. 44–76, and Kaplan, *The Ends of the Earth: A Journey at the Dawn of the 21st Century* (New York: Random House, 1996).

[8]This question is assessed in detail in Graham E. Fuller and Ian O. Lesser, *A Sense of Siege: The Geopolitics of Islam and the West* (Boulder, Colo.: Westview, 1995); see also Fred Halliday, *Islam and the Myth of Confrontation: Religion and Politics in the Middle East* (London: I. B. Tauris, 1996).

vating factor in the behavior of states, with or without an Islamist component. It is arguable that developments as disparate as the crisis in Algeria and the rise to power of Turkey's Refah Party have been driven as much by nationalism as Islam. Where Turkey's Muslim affinities are in tension with national security interests—as in relations with Syria—the nationalist impulse is likely to prove stronger. If the Spanish enclaves of Ceuta and Melilla emerge as a flashpoint in Spanish-Moroccan relations in the future, the driving force is likely to be Moroccan nationalism. So too, Egyptian nationalism will inevitably be a significant force behind Cairo's attitude toward issues affecting the Mediterranean and the Middle East as a whole.

The potentially destabilizing effects of nationalism are not limited to the Southern Mediterranean. The future security environments in the Balkans, the Aegean, and on Russia's southern periphery will be shaped by the strength of nationalist impulses. Indeed, the character of European policy toward the Mediterranean and the role of extra-Mediterranean powers, such as the United States, in Mediterranean security will be strongly influenced by the future balance between national and multilateral approaches. Growing unilateralism or the renationalization of foreign and security policies would surely complicate strategic dialogue and cooperation on Europe's southern periphery. Competing national visions can also influence the way in which the Alliance organizes itself to address security challenges on its periphery, as the political friction between the United States and France over control of the Southern Command in Naples (AFSOUTH) illustrates.

In a third trend, much discussion about the emerging strategic environment in the Mediterranean and the Middle East focuses on "low" (terrorism, political violence) and "high" (WMD) end threats. There is considerable merit in this approach, but it should not be allowed to obscure the continuing problem of the conventional defense of borders and the preservation of the territorial status quo. This problem—and the tendency to be distracted by other risks—is perhaps most acute in the Persian Gulf. But the Mediterranean basin also provides some important cases in which conventional clashes over territory and resources are possible. Prominent examples include altercations in the Western Sahara, Spain-Morocco (over the enclaves), Morocco-Algeria, Libya-Tunisia, Egypt-Sudan, Israel-Syria, the West Bank and Gaza, Greece-Turkey, and Turkey-Syria. This suggests that quite apart from the important potential for cooperation on counterterrorism and nonproliferation, the Mediterranean is a place where future demands for conventional peacekeeping, confidence-building measures, and security guarantees are likely to be high—a fact that has implications for how NATO structures its Mediterranean Initiative with the countries along the Southern Mediterranean littoral.

A fourth trend involves the end of Cold War alignments and the changing character of the Arab-Israeli dispute. These have opened the way for new security alignments and "geometries." Examples of this new fluidity in regional geopolitics include more overt Turkish-Israeli strategic cooperation, aimed largely at Syria, and the tendency of smaller Arab states, especially those in the Maghreb, to adopt an independent stance on security issues. Renewed progress in the Middle East peace process would facilitate strategic cooperation between Israel and Jordan, perhaps including Turkey in a trilateral alignment of status quo powers. In an extreme case, the advent of new Islamic regimes could drive secular but "revolutionary" regimes to make strategic agreements with the West, even if this requires a rapid disengagement with Israel.

Emerging links between Mediterranean nonmember states and NATO also suggest the possibility of a future in which European or Mediterranean institutions provide an alternative to security arrangements centered on the Middle East.[9] Whereas European security has an elaborate architecture, entailing multiple institutions—NATO, the Western European Union (WEU), and the Organization for Security and Cooperation in Europe (OSCE)—North Africa and the Middle East lack effective security institutions. In the Mediterranean setting, at least, some states may prefer to develop ties with existing European or Atlantic institutions based on a sense of affinity or the need for tangible security guarantees. This preference could provide an important incentive for non–NATO member Mediter-

---

[9]The recent experience of multilateral frameworks to address Middle Eastern security problems has been mixed, at best. See Bruce Jentleson, *The Middle East Arms Control and Regional Security (ACRS) Talks: Progress, Problems, and Prospects,* San Diego, Calif.: Institute on Global Conflict and Cooperation (IGCC) Policy Paper No. 26, September 1996.

ranean states to explore more active dialogue and cooperation with NATO, as well as the EU.

THE TRANSREGIONAL DIMENSION

Some of the most striking developments affecting the strategic outlook in the Mediterranean concern the steadily increasing interdependence of the European and Middle Eastern environments. In political, economic, and military terms, the future for both sides of the Mediterranean will be interwoven to a substantial degree.

On the political front, public and official opinion in North Africa and the Levant will be influenced by events in the Balkans and the Caucasus, as well as by events within Western European societies, that affect the position of Muslim communities. The Bosnian experience has been a watershed in this respect, and has served—rightly or wrongly—to confirm widespread suspicions in North Africa and elsewhere about European policy toward its Muslim periphery. In the Eastern Mediterranean, changes on the Turkish political scene may reinforce long-standing European perceptions of Turkey as a Middle Eastern rather than European state, complicating Turkey's desire to establish closer ties with European institutions. Indeed, as Europe continues to redefine itself in the wake of the Cold War, the perception of Turkish "otherness" is likely to grow. Yet, Turkey remains a member of the Atlantic Alliance, and risks on Turkey's borders will directly affect Turkey's European allies.

Even before the current stalemate—or perhaps reversal—in the Middle East peace process, European members of NATO had pressed for a greater role in Arab-Israeli negotiations, and Middle East diplomacy more generally. Lack of progress in the negotiations will tend to encourage even more-active European efforts in this direction, not least because Europe has a great deal at stake, both economically and in terms of stability on the periphery of the continent.[10] Similarly, much of the energy behind EU, NATO, and other initiatives toward North Africa and the Mediterranean has come from Southern European states with a special interest in North Africa

---

[10]See Gerald M. Steinberg, "European Security and the Middle East Peace Process," *Mediterranean Quarterly*, Winter 1996, pp. 65–80.

and a comparative advantage in North-South diplomacy. This is likely to be an important and continuing factor in shaping a European agenda that might otherwise be devoted almost entirely to challenges in Eastern and Central Europe.

In economic terms, there are many critical transregional linkages. Southern Mediterranean states recognize the extraordinarily important role of economic relations with the EU for their future prosperity, even if they are often uncomfortable with the reality of economic dependence.[11] The Euro-Mediterranean Partnership launched in Barcelona in November 1995 flows from this recognition, coupled with Europe's understanding of the need to foster development and stability across the Mediterranean. New lines of communication, including important new routes for energy transport, are another key point of interdependence. From the Western Mediterranean to the Caspian, the expansion of lines of communication for oil and gas is creating new opportunities for cooperation and conflict, with implications for the security and prosperity of both North and South. With new pipelines across the Maghreb and across the Mediterranean, and the potential for some part of future Caspian oil production to reach world markets via the Eastern Mediterranean (in addition to existing pipelines from Iraq to the Turkish coast), the Mediterranean region is becoming a focal point for energy trade and energy security concerns. Balkan reconstruction, and the revival of ports such as Thessaloniki and Trieste, would further reinforce the importance of the Mediterranean as a conduit for oil shipments from the Middle East to Eastern and Central Europe. Further afield, the opening of new transport links between Turkey, Iran, and Central Asia will offer the possibility of economic links to Europe via the Black Sea and the Mediterranean, rather than through Russia.

In hard security terms, the era of European sanctuary with regard to instability and conflict across the Mediterranean and beyond is rapidly drawing to a close. As the activities of Palestinian—and more recently Kurdish and Algerian—extremists demonstrate, European societies have long been exposed to the spillover effects of turmoil in

---

[11]See George Joffe, "Integration of Peripheral Dependence: The Dilemma Facing the Southern Mediterranean States," paper presented to Conference on Cooperation and Security in the Mediterranean After Barcelona, Mediterranean Academy of Diplomatic Studies, Malta, March 22–23, 1996.

North Africa and the Middle East.[12] In addition, Europe's greater Mediterranean periphery, from Algeria to Pakistan, displays a striking concentration of proliferation risks. The spread of WMD— nuclear, biological, and chemical—coupled with the proliferation of ballistic missile systems of steadily increasing range, is transforming the strategic environment around the Mediterranean. Southern Europe and Turkey will be the first within NATO to feel the existential effects of this exposure (major Turkish population centers are already within range of ballistic missiles deployed in Iraq, Iran and Syria), but not long after the year 2000, it is likely that every European capital will be within range of such systems.[13]

The mere existence of ballistic missile technology with ranges in excess of 1,000 km on world markets and available to proliferators around the Mediterranean basin would not necessarily pose serious strategic dilemmas for Europe. For the most part, the quest for regional prestige and "weight" is driving the acquisition of longer-range weapons, rather than the desire to hold European targets at risk. Given the diversity of frictions along South-South lines, it is likely that the Middle Eastern and North African neighbors of proliferators will face the first and most direct threat from weapons of mass destruction.[14] To date, many potential targets of WMD use around the Southern and Eastern Mediterranean have been reluctant to call attention to this risk, perhaps because WMD issues have tended to be seen as part of the strategic competition with Israel rather than a source of trouble in their own right. So too, countries such as Egypt and Israel have developed strong national agendas with regard to nuclear and other WMD matters. For this reason, proliferation questions will be difficult to treat in the context of NATO's Mediterranean Initiative, despite strong interest from NATO

---

[12]As the World Trade Center bombing demonstrated, the United States is also increasingly exposed to terrorism with roots in Middle Eastern problems.

[13]See Ian O. Lesser and Ashley J. Tellis, *Strategic Exposure: Proliferation Around the Mediterranean* (Santa Monica, Calif.: RAND, MR-742-A, 1996) and Yves Boyer et al., "Europe and the Challenge of Proliferation," *Chaillot Paper* No. 24 (WEU Institute for Security Studies, Paris, May 1996).

[14]The Iran-Iraq war, the civil war in Yemen, and the Gulf War provide examples along these lines. To date, the only concrete instance of ballistic missile attack against Western territory has been the ineffective Libyan Scud attack against Lampedusa in April 1986.

states. Again, progress on Arab-Israeli negotiations, including arms control, will be a key variable in determining the scope of Mediterranean efforts.

From a European perspective, the WMD and ballistic missile threat will acquire more serious dimensions where it is coupled with a proliferator's revolutionary orientation. Today, this is the case with regard to Iran, Iraq, Libya, and arguably Syria. But political circumstances could evolve in ways that would throw the WMD aspirations of other regional actors into sharper relief. Even short of dramatic changes in the political orientation of WMD-capable states, crises around the Mediterranean or in the Persian Gulf could raise the specter of WMD-related threats to European territory. Despite some initial concerns, risks from this quarter did not emerge during the Gulf War. But a future crisis involving Western intervention in the Middle East, if accompanied by more widespread WMD and ballistic missile capabilities, could end differently.[15]

As a result of proliferation trends, Europe will be increasingly exposed to the retaliatory consequences of U.S. and European actions around the Middle East and the Mediterranean basin, including the Balkans.[16] For the moment, conventionally armed ballistic missiles deployed on Europe's periphery are unlikely to possess the weight or accuracy to constitute a militarily significant threat. As a political threat and a weapon of terror capable of influencing the NATO decisionmaking during a crisis, their significance could be considerable. Would Southern European allies have offered the United States the same sort of access to facilities and military cooperation as they did during the Gulf War if their population centers were exposed to a credible threat of retaliation? Perhaps, but the deliberations would have been far more difficult, and the demands for defensive arrangements far more serious. As NATO begins to address the challenges of proliferation, and to the extent that the management of crises beyond Europe becomes a more prominent feature of European and transatlantic security cooperation, the Mediterranean and

---

[15]During the Gulf crisis, there was some concern that Iraq might have deployed ballistic missiles in Mauritania. There were also reports that Algeria may have accepted special nuclear materials transferred from Iraq.

[16]The possibility of Serbia acquiring improved Scud missiles capable of threatening Western Europe is discussed in Boyer et al., 1996, p. 12.

the potential role of a Mediterranean dialogue in containing proliferation risks will acquire additional significance.

THE EXTRA-REGIONAL DIMENSION

The consequences of trends in the Mediterranean security environment will reach well beyond Mediterranean shores. Under Cold War
conditions, the Mediterranean derived its primary strategic significance as an arena for competition between extra-Mediterranean su-
perpowers. The current environment has gone a considerable distance toward the visions of French (and many nonaligned) observers
who called for a "Mediterranean for the Mediterraneans." Russia has
withdrawn from the Mediterranean in security terms, although it re-
tains a stake in maritime access and Mediterranean political developments, and could play a more active role in the Balkans and on
Turkey's borders under certain circumstances. The United States
remains an overwhelmingly important military and diplomatic presence, especially in the Eastern Mediterranean. Challenges in the
Aegean, the Balkans, Turkey, and the Levant, not to mention the lo-
gistical tie to the Gulf, suggest that Washington's engagement in the
Mediterranean will be durable. To the extent that NATO devotes
more energy to the region, this too will tend to encourage a significant U.S. role. But European involvement in Mediterranean security
is substantial, and the critical economic and political relationships
between the North and South are, first and foremost, an EU responsibility. In this respect, the situation in the Mediterranean is quite
different from that in the Persian Gulf, where the United States plays
a dominant, and often unilateral, role as security manager.

In broad terms, the concerns of Mediterranean states, both North
and South, will be difficult to address without the engagement of key
non-Mediterranean states and wider European and Atlantic institutions. The range of hard and soft security issues characteristic of the
region—from proliferation to migration—favors multilateral approaches, and many of these issues would cause political discomfort
or be too costly to address unilaterally. An effective NATO role in
dialogue and security cooperation around the region will require a
minimum consensus on the importance of such a dialogue within
the Alliance. The growing recognition that insecurity across the
Mediterranean can have implications for *Europe as a whole*—not

simply Mediterranean Europe—is likely to encourage this consensus, as will the growing role of extra-European issues on the transatlantic security agenda.

Similarly, the EU's Euro-Mediterranean Partnership competes for aid and investment resources with demands from Central and Eastern Europe (and from Mediterranean Europe itself), and requires continuing support from non-Mediterranean EU members. As Germany moves toward a more forward-leaning approach to participation in military operations beyond its borders, contingencies on Europe's Mediterranean periphery may be the most likely setting for German involvement.[17] As key Central European states acquire a greater role in crisis management around the Mediterranean basin, there will be greater incentives to support policies aimed at longer-term stabilization and strategic dialogue.

Mediterranean security will also be influenced by actors beyond the European, Atlantic, and Eurasian spheres. The arms and technology transfer practices of China, North Korea, India, and Pakistan will have a bearing on the character and pace of WMD proliferation around the region. Anarchy and conflict in sub-Saharan Africa, Sudan, and the Horn of Africa could produce refugee crises affecting North Africa and Egypt, along with potentially destabilizing spillovers of political violence. If Europe is increasingly concerned about the risks emanating from the Southern Mediterranean, it should not be forgotten that states across the Mediterranean also face risks flowing from the even poorer and less stable regions to their "South."

KEY CONCLUSIONS

The foregoing analysis suggests a number of important conclusions:

The post–Cold War security challenges—broadly defined—are shifting from the center of Europe to the periphery, especially to the South. As a result, the Mediterranean is likely to become more important in the future in security terms. At the same time, traditional geographic distinctions are beginning to break down, and security problems in

---

[17]In this context, it is noteworthy that even during the Gulf War, a large part of the German Navy was deployed to the Mediterranean, replacing allied surface combatants transferred to the Gulf and the Indian Ocean.

the Mediterranean region are becoming increasingly interdependent.

Mediterranean security is, above all, a matter of internal security for states facing pressures for political, economic, and social change. These pressures will be especially pronounced in Mediterranean cities, where key political struggles (both violent and nonviolent) will be decided. In this context—and on both sides of the Mediterranean—questions of "personal security" and "security of identity" will play an important role in public opinion and policymaking. The EU is likely to be better placed than NATO to address these internal sources of insecurity.

Nationalism and the search for regional power and prestige will compete with Islamic politics as a key driver in the security future of the region. Substantial threats to the territorial status quo, driven by state-to-state frictions unburdened by Cold War constraints, will co-exist with threats from WMD or from the spillover of political violence. NATO will have a stake in the political-military consequences of these risks, and North-South dialogue can contribute to understanding and ameliorating them.

New security alignments are possible, even likely. These may take the form of a search for more favorable "geometries" on the part of actors around the Southern and Eastern Mediterranean who are interested in ties to a more stable European or Atlantic security order—that is, ties to NATO and the EU.

The transnational dimension of Mediterranean security is becoming more prominent as Europe and the Middle East become more interdependent in political, economic, and military terms. The growing European stake in the Middle East peace process, expanding lines of communication for energy through the Mediterranean and its hinterlands, and the steadily growing "reach" of weapons systems around the Mediterranean basin and beyond—all contribute to this trend. Europe will be increasingly exposed to the retaliatory and spillover consequences of developments on its Mediterranean and Middle Eastern periphery.

Extra-regional powers, above all the United States, will retain a strong stake and role in Mediterranean security. Hard and soft security challenges facing the region will be difficult or impossible to address without the engagement of non-Mediterranean states and European institutions. As tasks beyond Europe become a more central feature of transatlantic security arrangements, the Mediterranean will be a natural sphere for enhanced cooperation, with direct relevance to European stability.

IMPLICATIONS FOR THE ALLIANCE

These trends have several important implications for NATO. First, the Alliance faces an inevitable tension between its desire to address security risks emanating from the South (proliferation, terrorism, threats to friendly regimes) and its dialogue objectives with non–NATO member Mediterranean states. In theory, these aims are reconcilable, since enhanced dialogue should contribute to stability in the broadest sense. In practice, the dialogue partners (which include Egypt, Israel, Jordan, Mauritania, Morocco, and Tunisia) will view much of the content of NATO policy toward the South as directed "against them" (with the exception of Israel). At the same time, Southern Mediterranean states may wish to involve NATO more directly in their security problems—primarily through cooperation against internal security risks—and NATO members will have their own concerns about this involvement. Western policies with regard to political reform, human rights, and civil-military relations may severely constrain the scope for cooperation and may make dialogue on these issues difficult.

Second, although bilateral dialogue across the Mediterranean is useful, the real value of NATO's Mediterranean Initiative as a confidence-building device is only likely to be realized in a true multilateral format. As the preceding analysis suggests, most pressing security risks in the region are South-South, notwithstanding their implications for European and Atlantic security. Multilateral talks are necessary to address these problems directly, but the prospects for developing this approach are intimately tied to external variables, above all the Middle East peace process. NATO experience and expertise can play a role in facilitating progress in this sphere, but the Alliance cannot shape the outcome in the way that it might aspire to do in Eastern Europe, for example.

Finally, the very large role of EU policies in shaping the economic and political future of the Mediterranean region is a "permanently operating factor." Whatever the evolution of NATO's Mediterranean Initiative, it cannot be considered in isolation from the Barcelona process. To the extent that the complementary nature of NATO and EU initiatives can be made explicit, NATO's own activities in the region may win a better reception in Southern capitals. Moreover, the nature of Mediterranean security—on NATO's geographic periphery, but on Europe's doorstep—makes it an ideal subject for closer cooperation between European and Euro-Atlantic institutions.

Chapter Two

THE BARCELONA PROCESS AND OTHER
MEDITERRANEAN INITIATIVES

During the last decade a number of initiatives have been launched or proposed to enhance cooperation in the Mediterranean: the "Five plus Five," the Conference for Security and Cooperation in the Mediterranean (CSCM), the Mediterranean Forum, Middle East and North Africa (MENA) Summits, the Arms Control and Regional Security Working Group (ACRS), the European Union's Euro-Mediterranean Partnership (the Barcelona process), the WEU Mediterranean Initiative, and the OSCE's Mediterranean Contact Group.

The multiplicity of frameworks dealing with Mediterranean security raises several basic questions for the Alliance:

- How does NATO's Mediterranean Initiative relate to other Mediterranean endeavors?

- What is NATO's comparative advantage and specific contribution to Mediterranean security?

- How can NATO's initiative be harmonized with those of other institutions in a way that enhances overall security in the Mediterranean and avoids duplication of effort?

To answer these questions it is useful to look in some detail at the other initiatives. This may make it easier to assess how NATO's initiative fits into the broader pattern of cooperation in the Mediterranean and what its specific contribution to Mediterranean security should be.

THE EURO-MEDITERRANEAN PARTNERSHIP

The most ambitious and developed Mediterranean initiative is the Euro-Mediterranean Partnership (or the Barcelona process, as it is commonly known). The initiative responded to a perceived need, particularly among the countries in Southern Europe, to address more forthrightly the growing social and economic problems on the non-European side of the Mediterranean littoral. It also reflected a desire among Southern EU members to redress a perceived imbalance in the allocation of resources between the East and the South.

The Euro-Mediterranean Partnership, however, did not emerge in a vacuum. Rather it was the culmination of a decades-long series of European efforts to deepen cooperation with the countries of the Southern Mediterranean littoral. The first contacts between the European Economic Community (EEC) and Mediterranean non–ECC members date back to the 1960s. In their initial form, contractual links were limited to trade and provided for the free access of industrial goods to EEC markets and for specific agricultural export concessions. However, the establishment of these accords had more to do with Southern Mediterranean countries' desire to increase exports to Europe than with a comprehensive European cooperation strategy for the region.

These links were extended in the 1970s to include economic and financial ties in the form of association agreements with Turkey, Cyprus, and Malta and of cooperation agreements with other Mediterranean countries. Euro-Mediterranean trade relations experienced a downturn during the 1970s and early 1980s as a result of the oil crisis and the admission of Spain and Portugal into the EEC. During this period, the EEC established restrictions on sensitive products, notably textiles and agricultural goods. Exports from Southern Mediterranean states to the community remained at traditionally low levels.[1]

The end of the Cold War provided the opportunity to develop more-ambitious programs, and a new Euro-Mediterranean policy was introduced by the European Council in 1990. Backed by increased fi-

---

[1]"The European Union's Relations with the Mediterranean," European Commission Memorandum 94/74, December 1994.

nancial support, improved market access and an effort at political dialogue, the policy was based on the assumption that European Community (EC) assistance coupled with structural reform would stimulate private-sector activity in Southern Mediterranean economies and lead to long-term growth.

Despite gradual improvements, the efficacy of the pre–Barcelona process strategy was limited. The European Community had little effect on improving the performance of Southern Mediterranean economies. The high degree of external protection in most partner economies persisted, while European countries remained reluctant to open their agricultural product markets. The lack of economic liberalization prevented regional exports from diversifying and offsetting the major structural deficit in trade these countries have historically suffered vis-à-vis the EC.[2] The political dialogue foreseen by the 1990 Euro-Mediterranean policy review acquired substance, partly as a result of the failure of the Arab Maghreb Union (AMU) and continuing Arab-Israeli tensions.

More fundamentally, the major flaw of pre–Barcelona process policies was the lack of a strategic vision of the Mediterranean as a single geopolitical entity; in fact, up until the early 1990s, the EC was crafting distinct approaches for the Maghreb, Mashreq, and Israel. It was against this backdrop of slowly evolving economic cooperation and lackluster results that the European Union began planning for a substantial overhaul of its Mediterranean policy, which culminated in the 1995 Barcelona Conference.

The 1995 Barcelona Conference

On November 27–28, 1995, the European Union formally launched its new Mediterranean policy in Barcelona. The Barcelona Conference represented the pinnacle of a year-long effort by the EU to lay a stronger basis of partnership with Algeria, Cyprus, Egypt, Israel, Jordan, Lebanon, Malta, Morocco, Syria, Tunisia, Turkey, and the Palestinian National Authority. The qualitative change in EU Mediterranean policy was mainly promoted by France, Italy, and

---

[2]"The European Union's Mediterranean Policy," European Commission Memorandum 94/63, October 1994.

Spain during the Corfu and Essen summits of the European Council in 1994.

The main aim of the Euro-Mediterranean Partnership is to promote long-term stability through economic development and liberalization. Economic development is assumed to have positive spillovers in political, social, and security terms by providing more opportunities for jobs at home (thus easing migratory pressures), raising the standards of living, and decreasing the attractiveness of extremist ideologies. Liberal economic policies are in the long term also expected to lead to more liberal democratic institutions, which will have the effect of enhancing cooperation and stability both domestically and internationally. The initiative is designed to promote an integrated economic area that fosters closer cooperation on political, social, and economic issues. The most noteworthy aspect of the Euro-Mediterranean Partnership is its long-term and comprehensive approach, which was absent from previous European efforts in the region.

The 27 delegations at the Barcelona Conference agreed on a declaration of principles and a work program. The documents call for the establishment of a truly comprehensive partnership—the Euro-Mediterranean Partnership—among the participants in three different baskets:

- economic and financial cooperation

- social and cultural affairs cooperation

- political and security partnership.[3]

The Euro-Mediterranean Partnership entails a multilateral political, economic, and social dialogue between the EU and its 12 Mediterranean partners; strengthened cooperation between the civil societies of the participating countries; and a series of association agreements. To support economic development, the EU supplements the agreements with financial assistance—primarily through the MEDA grant program—totaling ECU 4.7 billion ($6.1 billion) over

---

[3]"Barcelona Declaration and Work Program," issued at the Euro-Mediterranean Conference, Barcelona, November 27–28, 1995.

the 1995–1999 period and an equal amount of funds provided as loans by the European Investment Bank.[4]

Through progressive elimination of tariff and nontariff barriers to trade, the association agreements should lead to a Euro-Mediterranean free-market area by 2010, in compliance with World Trade Organization guidelines. The Euro-Mediterranean free-trade area will provide for reciprocal free trade of all manufactured products between the EU and most Mediterranean countries, for preferential and reciprocal access for some agricultural products, and for free trade among partner countries. The EC expects the Euro-Mediterranean zone to constitute the largest free-trade area in the world, covering 600 to 800 million people and some 30 to 40 countries. The creation of a free-trade area should streamline the partners' regulatory and economic policy frameworks, raise their long-term competitiveness, attract substantially more private direct investment, improve mutual understanding and awareness, and accelerate sustainable economic and social development.[5]

Progress Since Barcelona

The Barcelona Conference set up a series of agreements, meetings, and activities in the three baskets. Progress has varied depending on the particular basket and issue:

Economic and Financial Cooperation. The Euro-Mediterranean policy's most significant results to date are in the economic domain. Four association agreements have so far been signed—with Tunisia, Israel, Morocco, and Jordan, respectively; the EU has also initialed a similar agreement with the Palestinian Authority.[6] Association negotiations with Egypt are likely to be completed by early 1998, while Lebanon is expected to sign an agreement with the EU later in 1998. Talks with Algeria have been slowed by opposition from domestic

---

[4]"The Barcelona Conference and the Euro-Mediterranean Association Agreements," European Commission Information Memorandum, 23 November, 1995.

[5]"The Commission Proposes the Establishment of a Euro-Mediterranean Partnership," *European Commission Press Release 94/56*, October 1994.

[6]Cyprus and Malta signed association agreements in the 1970s; Turkey has a Customs Union with the EU, which went into effect on January 1, 1996.

trade unions and internal civil strife; Syria has embarked on exploratory talks with the EU but progress is unlikely in the foreseeable future. EU bargaining with Lebanon and Egypt has been particularly slow and controversial as a result of mutual reluctance to compromise on agricultural export quotas and the pace of trade liberalization.[7] Regarding negotiations with Morocco, Tunisia, and Egypt, access to the European market for these countries' agricultural and agro-industrial products has also proven to be a major stumbling block.[8]

Association agreements are expected to promote trade among Southern Mediterranean countries, so that North-South cooperation will be reinforced by stronger South-South economic relations. In 1995, the value of intra-regional trade in the Southern Mediterranean was only 5 percent of the area's external trade.[9]

Despite recent progress, certain political-economic issues will be difficult to resolve. For instance, the EU's economic strategy for the Mediterranean promises long-term benefits and imposes significant short-term social and economic costs for partner countries. This, coupled with the dismal macroeconomic performance by the Middle East and North Africa region in recent years,[10] is likely to discourage Mediterranean partners from undertaking painful economic reform. Moreover, the contrast between the economic challenges facing these states and the relatively modest amount of funds allocated by, for example, Brussels is considerable.

According to EU plans, official EU funding should eventually be dwarfed by flows of private capital to partner countries; yet the Barcelona process cannot by itself create the solid institutional base

---

[7] "Mideast Overshadows Malta Talks," *Financial Times*, April 16, 1997.

[8] For instance, the association agreement with Morocco encountered German opposition over an import provision for 5,000 tons of Moroccan cut flowers, worth about $2 million (personal interview with EC official, November 25, 1995).

[9] "Strengthening the Mediterranean Policy of the European Union," *Bulletin of the European Union*, Supplement 2/95, p. 17.

[10] In a recent study, the World Bank concluded that the region's weak economic performance is related to flat trade ratios, high tariffs, a decrease in foreign direct investment, and the fact that all countries, except Tunisia, are not investment grade. See "Middle East and North Africa," *Global Economic Prospects and the Developing Countries*, (Washington, D.C.: The World Bank, 1996).

and political stability necessary to attract foreign investment in Southern Mediterranean countries. European investors generally avoid North Africa and the Middle East because of the region's local institutional base, ambiguous legal framework, and bureaucratic obstacles. The EU needs to play a stronger role in improving governance in Mediterranean countries. However, the key decisions have to be made by the countries themselves.[11]

Cooperation in Social, Cultural, and Human Affairs. The preservation of architectural and intellectual heritage—the theme of the first Euro-Mediterranean meeting of culture ministers—has been the most successful subject of cooperation in this basket. Negotiations about other matters—such as drug trafficking, terrorism, organized crime, and immigration—have been slow. Some partner countries, such as Egypt, believe the EU is discriminating against them because the harsher repatriation rules recently approved by the European Council will appear in their association agreements but will not be retroactive, so that countries that have already signed association agreements, such as Morocco, will continue to enjoy a more lenient set of regulations.[12] The issue of human rights has also been a source of contention at EU meetings with Egypt and Lebanon.[13]

The sensitivity of Europe discussing terrorism, drug trafficking, and immigration with Arab states has slowed down related negotiations considerably.[14] There is a widespread belief in partner countries that

---

[11]In a recent study, Bernard Hoekman and Simeon Djankov note that association agreements do "little to ensure investors of national treatment or to grant the general right of establishment," and therefore stifle foreign direct investment (FDI) flows into partner countries. FDI is essential in fostering economic development because it usually targets key areas of the economy, such as telecommunications, information technology, port services, financial intermediation and business support services. See Bernard Hoekman and Simeon Djankov, "The European Union's Mediterranean Free Trade Initiative," *The World Economy*, Vol. 19, No. 4, July 1996, p. 403.

[12]"Arabs to Dutch Minister: EU Overemphasizing Security," Rotterdam *NRC Handelsblad*, April 2, 1997: FBIS-WEU-97-094.

[13]Egypt has accused the EU of not being consistent in its support of human rights, arguing that Brussels is not vocal enough in demanding the respect of the self-determination and economic rights of Palestinians in Israel. See "Egypt's Musa on Mideast Peace, Other Topics," Cairo *MENA*, April 15, 1997: FBIS-NES-97-073.

[14]The approval of the final draft of a declaration was held up by disagreements over the definition of terrorism, with Lebanon wanting to clarify the right to use force in the liberation of the occupied territories. There were also disagreements over whether

the EU is more interested in stemming the flux of economic migrants, terrorists, and drugs crossing the Mediterranean to Europe than in promoting the region's economic development. Such skepticism has effectively prevented a breakthrough in discussions of sensitive issues and is likely to prevent substantive progress in the future unless mutual trust is generated.

Political and Security Partnership. Cooperation in political and security affairs is coordinated by a committee of senior government officials from European and partner countries. In principle, security cooperation is designed to foster good neighborly relations by establishing early warning systems, organizing Euro-Mediterranean meetings of senior military officers, sharing military expenditure data, promoting exchanges of visits at military academies, and training arms control experts from partner countries.

A fax network connecting the EU with the 12 Mediterranean partners has been set up and was used to compile a list of international instruments concerning human rights to which partner countries have acceded. This will be complemented by an electronic information network that links the foreign ministries of all participating countries.

As a way of integrating civil society cooperation with its own Euro-Mediterranean security partnership, the European Commission is supporting a network of international affairs and strategic studies institutes (EuroMeSCo). EuroMeSCo has created two working groups, one on confidence-building, conflict prevention, arms control, and disarmament and a second on political and security cooperation.[15]

The Euro-Mediterranean dialogue on security matters has also been complicated by the deterioration of Arab-Israeli relations since the election of Israeli Prime Minister Benjamin Netanyahu in May 1996. As a result, the Arab countries, especially Syria and Lebanon, have opposed military contacts or activities that would include Israel. The Malta summit of foreign ministers on April 15–16, 1997, failed to ap-

---

partner states had an *obligation* or *responsibility* for readmission of migrant workers expelled from Europe. The parties agreed to use the term *responsibility*.

[15]See Roberto Aliboni, Abdel Monem Said Aly, and Alvaro de Vasconcelos, *Joint Report of the EuroMeSCo Working Groups on Political and Security Cooperation and Arms Control, Confidence-Building and Conflict Prevention*, April 1997.

prove a number of confidence-building measures worked out by senior officials during the previous months. The idea of setting up a conflict prevention center was put on ice, as was a proposal to establish a network of defense institutes. A similar fate befell the French and Maltese proposal for a stability pact in the Mediterranean designed to put into effect the principles and goals contained in the Barcelona declaration.[16]

The difficulties surrounding the dialogue on security matters evident at the Malta summit underscore the linkage between the Middle East peace process and Euro-Mediterranean security cooperation. While the fact that all partner countries decided to attend the Malta summit during a period of deteriorating Arab-Israeli relations illustrates the importance they attach to the Barcelona process, the near collapse of the peace process effectively eliminated any chance of reaching consensus on such controversial issues as terrorism or immigration. Most of the talks among partners took place bilaterally to defuse the Har Homa crisis, and little attention was paid to important agenda items such as the stability pact.

While EU officials stress that the Barcelona process is not a substitute for the Middle East peace process, EU relations with Syria, Lebanon, Israel, and the Palestinian National Authority will be inevitably influenced by the state of Arab-Israeli relations. The preliminary talks under way with Syria, for instance, are not likely to lead to the signing of an association agreement unless the peace process makes substantial progress. In fact, Syria's presence is a wild card for the political and security dialogue of the Barcelona process. The participation of Damascus in EU-sponsored security discussions is significant, since Syria has generally been reluctant to join multilateral initiatives in the region. Its involvement was made more palatable by the fact that the large membership of the Barcelona process provides a certain political "cover." Nonetheless, Syria has thus far been a reluctant partner in security talks, opposing the involvement of military officials. Unless Syria and Israel improve their relations, the necessary consensus for more substantive security cooperation will be difficult to achieve.

---

[16]See Fred Tanner, "Die Mittelmeer-Partnerschaft der Europäischen Union in der Krise," *Neue Zürcher Zeitung,* July 8, 1997.

The European Union also faces the challenge of developing what it terms "mechanisms for the peaceful resolution of disputes" and "early warning systems" without encountering the same obstacles similar efforts have faced in the context of the peace process. The experience of the Arms Control and Regional Security Working Group (ACRS—see below) is particularly instructive. The ACRS successfully negotiated a declaration of principles on confidence-building measures and designed a series of practical measures to enhance regional stability. Despite the early success, the implementation of such agreements has yet to materialize because of Egyptian insistence that Israel sign the Nonproliferation Treaty (NPT). It will be difficult for the EU to keep even modest confidence-building measures from becoming hostages of the nuclear issue. The EU should also be concerned about the view expressed by some Arab countries that Europe is overemphasizing political security in the Mediterranean at the expense of encouraging economic progress.

The EU's Future Role in the Mediterranean

One of the major predicaments of crafting a comprehensive diplomatic approach for the Mediterranean is how to connect such a large and heterogeneous group of countries in a structure of cooperation. Through the use of a loose institutional framework, the European Union has developed what is likely to be an effective strategy. The Barcelona process permits differential treatment while at the same time encouraging gradual harmonization. Despite its sound approach, the EU will have to contend with influential events and trends, in all three baskets listed below, that lie outside its direct control.

Economic Development. Economic development in Mediterranean countries depends critically on organizing efficient public institutions, domestic competition, a well-functioning service sector, investment in human resources, high rates of private saving and investment, and a stable economy. Most of these cannot be "exported" from the EU and require politically unpalatable decisions on the part of partner countries. The gradual approach to economic reform may also be too slow to enable Mediterranean partners to compete with Eastern European and Asian economies—which in turn are driven by liberalization measures achieved during the General Agreement of Tariffs and Trade (GATT) Uruguay Round.

Democratization. EU influence on democratization and other social issues is tenuous at best and likely to bear fruit only in the long run. Moreover, the Barcelona process is not equipped to deal with the consequences of social instability spurred by forces such as Islamic extremism. The civil strife in Algeria has seen minimal EU involvement and is a prime example of how the divergence of views between France, other EU members, and the United States hampered Western efforts to defuse the crisis.

The Middle East Peace Process. As was apparent at the foreign ministers' summit in Malta in April 1997, the Middle East peace process can greatly affect the pace of the Euro-Mediterranean security and political dialogue. The peace process has also exposed the absence of a clear division of labor among the United States, the European Union, and other institutions in regard to the region. These issues are important because they imply that the Euro-Mediterranean security partnership is likely to be affected by the agendas of other bilateral and multilateral efforts toward Mediterranean cooperation. The next section examines these initiatives in greater detail.

OTHER MEDITERRANEAN COOPERATION INITIATIVES

In addition to the Barcelona process and NATO's Mediterranean Initiative, several other Mediterranean cooperation initiatives exist—some were sponsored by a number of states, others by multilateral institutions. The earliest post–Cold War efforts were crafted by groups of states on the Mediterranean littoral—the "Five plus Five" and the CSCM, for instance, were first floated in 1990. Initiatives involving several countries were also devised in the field of arms control, with the work of ACRS being the most significant. More recently, a series of U.S.-sponsored conferences—called MENA economic summits—have been held yearly since 1994 to promote investment in the region.

The first multilateral institution to set up a Mediterranean dialogue was the WEU in 1992. The WEU was followed by the OSCE, NATO and the EU's Barcelona process in later years. These organizations have become involved in the region to complement state-sponsored initiatives as well as to compensate for the lack of success of some early efforts, such as the "Five plus Five" and the CSCM.

The Conference on Security and Cooperation in the Mediterranean (CSCM)

In 1990, Italy and Spain proposed the creation of CSCM as a Southern counterpart to the Conference on Security and Cooperation in Europe (CSCE). This framework was designed to mirror the three-basket approach of the Helsinki process and imitate CSCE decisionmaking mechanisms. The CSCM faced several important obstacles and was never officially approved. The first obstacle was the difficulty of conducting talks with states from a vast and diverse area (membership would have been open to countries from Southern Europe, the Balkans, the Middle East, and the Persian Gulf). The second obstacle was American fears that the CSCM would derail the Middle East peace process and curtail U.S. freedom of action in the Mediterranean. The third obstacle was the complex nature of the planned CSCM decisionmaking process, which was considered to be overly burdensome. Although the CSCM as a framework is not likely to be resuscitated, its principle of linking military security to an overall strategy of cooperation and partnership has been enshrined in the Barcelona process and ACRS.

The "Five plus Five"

The creation of the "Five plus Five" in July 1990—originally established as the "Four plus Five" and later enlarged to include Malta later in 1990—was aimed at developing cooperation between Spain, France, Italy, Portugal, and Malta, on the one hand, and Algeria, Libya, Morocco, Tunisia, and Mauritania on the other. Discussion topics included natural resource management, economic links and financial assistance, immigration, and culture. The "Five plus Five" very soon encountered serious obstacles, the most important of which was the international embargo on Libya. The absence of Egypt, friction between Algeria and Morocco over the Western Sahara and Algeria's internal instability also undermined the initiative.

The "Five plus Five" excluded military security discussions from the agenda; in that regard it presented a unique approach and may have generated greater interest in Arab countries keen to stimulate their economic development. The fundamental flaws of the "Five plus Five" lay in its "convoy" approach to decisionmaking—which allowed the pace of the discussions to be determined by the most reluctant partner—and the related problems of Libya's pariah status.

The Forum for Dialogue and Cooperation in the Mediterranean

Having been excluded from the "Five plus Five," Egypt, in 1991, proposed the organization of the Forum for Dialogue and Cooperation in the Mediterranean, or Mediterranean Forum. France cosponsored the initiative, and subsequently the Mediterranean Forum won the endorsement of Spain, Italy, Greece, and Portugal before being launched in Alexandria in July 1994. In Alexandria, the 10 founding members—Algeria, Egypt, France, Greece, Italy, Morocco, Portugal, Spain, Tunisia, and Turkey—admitted Malta as an 11th member.

The participating states set up three working groups on political dialogue, dialogue between cultures and civilizations, and economic and social cooperation. Unlike the "Five plus Five," the Mediterranean Forum focuses in part on security issues.

The major problem with the Mediterranean Forum is that most of its functions are being overtaken by the Barcelona process. This need not imply the end of the forum's usefulness, as long as member states find a way of differentiating its activities from the EU's framework. The Mediterranean Forum, for instance, could exploit the special relationship North African countries have with Southern Europe to discuss issues that are too sensitive for high-profile multilateral discussion. At the May 1996 Mediterranean Forum summit, it was suggested that the forum could become an informal framework on matters of common interest. Given that the countries most closely involved in the peace process (notably Israel, Lebanon, and Syria) are not forum members, this initiative could serve as a caucus of Mediterranean states that can informally discuss topics still regarded as off-limits in the Barcelona process (such as WMD proliferation and terrorism) without becoming entangled in disputes related to ongoing regional conflicts.

The Arms Control and Regional Security Working Group (ACRS)

ACRS was established during the Madrid Peace Conference of 1991 as part of the multilateral track of the peace process negotiations and is composed of Israel and 12 Arab countries (with the notable exception of Syria, Libya, Iraq, and Iran). Work is divided into conceptual and operational baskets, and progress in each has been considerable. Negotiations in 1993 and 1994 produced a first draft of a declaration of principles of peace and security in the Middle East, modeled on the 1975 Helsinki Final Act. Progress in the operational basket focused on proposals for joint rescue-at-sea exercises, the establishment of a regional security center in Amman, and a regional communication center in Egypt. The last meeting of ACRS took place in Tunis in December 1994 and failed to approve the declaration of principles and specific items in the operational basket.

The lack of a formal document belies the actual progress made in ACRS discussions; the flurry of activities discussed in this forum was unprecedented for the Mediterranean and Middle East. Moreover, the mere fact that Israeli and Arab officials interacted and exchanged views on security issues was in and of itself a confidence-building measure.

However, ACRS depends on voluntary and, therefore, easily revocable commitments. Several key countries, such as Syria, Lebanon, Libya, Iraq, and Iran, have refused to join ACRS; Syria and Lebanon will continue to remain outside this framework until a breakthrough in their bilateral negotiations with Israel. The most significant problem, however, has been the disagreement over the nuclear issue, which has led to an impasse in the talks.[17]

The Middle East and North Africa (MENA) Economic Summits

The World Economic Forum and the Council on Foreign Relations jointly organized in October 1994 a Middle East/North Africa eco-

---

[17]For a detailed discussion, see Shai Feldman, *Nuclear Weapons and Arms Control in the Middle East* (Cambridge, Mass.: MIT Press, 1997), pp. 8–15.

nomic conference in Casablanca. The summit's objective was to create partnerships between the public and private sectors and focus on immediate commercial opportunities and projects in the Middle East and North Africa. MENA summits have taken place yearly since that time. The creation of the MENA Development Bank (MENADB), one of the key institutions related to this initiative, was approved during the 1995 Amman summit and is expected to begin operations in Cairo in 1998. However, the stalemate in the Arab-Israeli peace process has had a spillover effect on the MENA process. Many Arab countries—including Algeria, Egypt, Lebanon, Morocco, Saudi Arabia, and Syria—refused to send representatives to the 1997 MENA summit in Doha, Qatar, in protest of the policies of the Netanyahu government.

Though MENA conferences are viewed by several participants as too closely associated with U.S. policy, the meetings contribute to making the Mediterranean an economically coherent and integrated region. The MENA framework would encounter major opposition from European states if it sought a more substantive role in shaping regional economic policy. EU countries fear that the MENADB's activities would be dominated by the United States and would be biased toward Middle East projects to the detriment of North African investment needs. As a result, EU states including Germany, France, and the United Kingdom have maintained that they will not participate in the MENADB.[18] The failure of the U.S. Congress to authorize the transfer of $52 million to the MENADB in 1997 has also raised serious questions about the bank's future.

The WEU Mediterranean Framework

The WEU is also involved in Mediterranean cooperation. In the Petersberg Declaration of June 1992, the WEU Council entrusted the Presidency and Secretariat to develop ties with Algeria, Morocco, and Tunisia. The discussions—later extended to Egypt, Mauritania, and

---

[18]European countries, however, have agreed to continue negotiations on the EU-backed Middle East and North African Intermediation Organization, which would include a forum for regional cooperation and provide a framework for coordinating economic policies and identifying regional projects funded by existing institutions. See Leon T. Hadar, "Meddling in the Middle East?" *Mediterranean Quarterly*, Fall 1996, p. 52.

Israel—have as their main purpose the exchange of views on security and defense issues affecting the Mediterranean region. Given the limited agenda of the WEU framework, the results have been very modest—limited to occasional expert meetings and diplomatic contacts between the WEU and the Brussels embassies of the partner countries.

WEU Mediterranean policies have received increased attention as a result of the creation of EUROFOR and EUROMARFOR. The missions earmarked for the two forces—composed of Spanish, Italian, French, and Portuguese units—include such humanitarian functions as the emergency evacuation of European citizens in the event of a crisis. North African countries viewed the creation of these forces with considerable suspicion and were quick to voice their objections in November 1996, when the EUROFOR headquarters opened in Florence. Although Libya took the lead in sharply criticizing the establishment of these WEU task forces, other countries, such as Tunisia, felt compelled to show solidarity with Tripoli. One of the main problems was the lack of adequate information and openness about the goals and purpose of EUROFOR and EUROMARFOR. Some of the concerns expressed by Arab countries were allayed by inviting their experts to observe WEU military exercises and to visit the EUROFOR headquarters in Florence. Nonetheless, discussion partners continue to express dissatisfaction about the way these multinational forces were set up and about the perceived lack of clarity over their missions.

The sharp Arab reaction to the establishment of EUROFOR and EUROMARFOR underscores the importance of providing adequate information and properly laying the groundwork for Western initiatives in the region. This is of particular relevance to NATO, since the Alliance has an even stronger image problem in the Mediterranean. Both NATO and the WEU need to ensure that their activities are correctly perceived by Southern Mediterranean publics to prevent good-faith initiatives from backfiring.

The OSCE's Mediterranean Contact Group

The OSCE began its Mediterranean outreach program at the 1994 Budapest Review Conference. Italy, Spain, and France took the lead in proposing that the OSCE establish an informal contact group with experts from Algeria, Egypt, Israel, Morocco, and Tunisia. As part of the OSCE effort, a seminar was organized in Cairo in 1995 on OSCE's experience in the field of confidence-building measures; a recent decision should allow the Forum for Security Cooperation (an OSCE body) to share its findings with Mediterranean partner states.[19]

Mediterranean partners can benefit from learning how OSCE operates—the activities and routines provide useful models of how confidence-building measures can be implemented in their region. In fact, OSCE mechanisms have been used in several Mediterranean initiatives, such as ACRS and the Barcelona process. These more capable and regionally focused frameworks, however, will most likely limit OSCE's direct role in the Mediterranean to information sharing and expert meetings.

COMPLEMENTARITY AND COMPETITION

At present, there are seven ongoing Mediterranean cooperation initiatives: NATO's Mediterranean Initiative, EU's Euro-Mediterranean Partnership (the Barcelona process), WEU's Mediterranean Initiative, OSCE's Mediterranean Contact Group, the Mediterranean Forum, ACRS, and MENA conferences.[20] Several of these programs closely overlap and may in some cases undercut each other's effectiveness. The myriad of initiatives raises the question of how the frameworks relate to one another and how NATO can maximize its effectiveness in such a crowded "policy market."[21]

Table 1 compares the various Mediterranean efforts in terms of their goals, assumptions, scope, resources, partner countries, and prospects.

---

[19]"The 1996 Lisbon Document," issued at the OSCE Summit of Heads of Government, Lisbon, December 2–3, 1996.

[20]The "Five plus Five" and CSCM are not viable frameworks and will not be discussed further.

[21]For a broader discussion of the problem of complementarity and competition among various Mediterranean frameworks, see Roberto Aliboni, "Institutionalizing Mediterranean Relations: Complementarity and Competition," *Internationale Politik und Gesellschaft*, Nr. 3, 1995, pp. 90–99.

Table 1

Summary of Mediterranean Cooperation Initiatives

Initiative	Goal	Assumptions	Scope	Re-sources	Mem-bership	Pros-pects
NATO	Improve own image; increase mutual under-standing, mili-tary cooperation	NATO has a bad image in Arab world; reputation improved with dialogue and some military cooperation	Security; info. ex-change and some practical coop-eration	Limited	Limited (6); does not in-clude Syria, Algeria	Fair
EU	Stability in the Mediterranean to prevent threats and develop new markets	Stability through economic inte-gration, security, and cultural dia-logue	Economic; social; security with no military di-mension	Sub-stantial	Largest (15); with Syria, Al-geria	Good
WEU	Improve own image; increase mutual under-standing	WEU has a bad image in Arab world; reputation improved with dialogue	Security; info. ex-changes	Very lim-ited	Limited (6); includes Algeria, not Syria	Poor
OSCE	Extend OSCE experience to Mediterranean states	Exposure to OSCE work facilitates CBMs in region	Security; info. ex-changes	Very lim-ited	Limited (5); includes Algeria, not Syria	Poor
Mediter-ranean Forum	Informal meet-ings to discuss common inter-ests	Group is willing to discuss sensitive issues	Economic; security; social	Very lim-ited	Limited (5); includes Algeria, not Syria	Fair/ Poor
ACRS	Multilateral fo-rum to address arms control; CBMs	Enduring peace depends on re-gional arms control and CBMs	Security; info. ex-changes; and some practical coop-eration	Limited	Large (12); does not include Syria	Fair/ Poor
MENA	Create invest-ment opportu-nities	Stability depends on economic in-tegration	Economic; annual summits	Limited	Large, with country and business reps	Fair/ Poor

With the exception of the MENA summits, all these initiatives include some discussion of security aspects, reflecting the importance European countries attach to such issues. Of the security initiatives, NATO's is the one with the greatest potential to develop practical cooperation and military links. Economics is the second most common field of cooperation. While the EU's Barcelona process is by far the most extensive attempt at a comprehensive dialogue, no single initiative can claim to cover the entire spectrum of cooperation when security issues cover both soft confidence-building measures *and* concrete military measures.

NATO, the EU, the WEU, and the OSCE have to balance their Mediterranean dialogues with other more pressing European security priorities. Mediterranean cooperation is only a peripheral concern for the OSCE, a secondary concern for NATO and the WEU, and a moderately significant one for the EU. The relatively low importance attached to Mediterranean cooperation by multilateral institutions is a principal reason why, with the exception of the EU, resources to sustain these initiatives are likely to remain limited.

The table also shows that the NATO and WEU programs have similar goals. However, NATO's prospects for success are better than the WEU's because NATO is institutionally more capable than the WEU to pursue an outreach strategy based on military cooperation. While NATO resources are limited, the Alliance can rely on the unique experience gained from the North Atlantic Cooperation Council (NACC) and the Partnership for Peace (PfP).

The limited membership of NATO's Mediterranean initiative may also be an advantage. Syria and Lebanon, for instance, have been reluctant to discuss security and military issues within the EU-sponsored Barcelona process while the bloody civil war in Algeria has created a disinclination on the part of many Western countries to strengthen ties with the Algiers regime. NATO enjoys the advantage of not having these countries as participants in its dialogue. This makes it easier to discuss security issues, and this limits—though does not eliminate entirely—the effect of the Middle East peace process on the dialogue.

COOPERATION AND COORDINATION

Given the multitude of initiatives that already exist in the Mediterranean, there is a need for closer coordination of efforts, particularly between Western initiatives, to ensure that they complement one another and don't work at cross purposes. Here the goal should be to achieve synergy.

Greatest synergy is achieved when strategic goals converge and competitive goals diverge—when activities and skills are complementary. The potential for competitive conflict is minimized when each initiative brings a distinctive perspective to Mediterranean cooperation. For instance, NATO could decrease the potential for competitive conflict with the EU by focusing more on military cooperation programs—a field where the Alliance has a significant comparative advantage over the EU. Whether having similar membership helps or hurts coordination among initiatives depends on how similar the frameworks are in their activities; similar activities and membership make harmonization more difficult.

Figure 1 illustrates the potential for synergy of the NATO Mediterranean Initiative with its counterparts. While all initiatives share the same overarching goal, they differ in their degree of complementarity, interdependence, and prospects for success vis-à-vis the NATO dialogue. The efforts least likely to produce synergy with NATO's are MENA, ACRS, and OSCE. Their tasks are not very complementary

RANDMR957-1

	Complementarity in Tasks/Members	Interdependence	Prospects for Success	Gains from Coordination
EU	High	High	High	High
WEU	Moderate	High	Low	Moderate
OSCE	Moderate	Low	Low	Moderate
Mediterranean Forum	Moderate	Moderate	Low	Moderate
ACRS	Low	Moderate	Moderate	Moderate
MENA	Low	Low	Moderate	Moderate

■ High   ▨ Moderate   □ Low

Figure 1—Mediterranean Initiatives and Their Potential for Cooperation with NATO's Initiative

with NATO's, and their prospects for success range from moderate to low.

The Mediterranean Forum is partly complementary since it might involve informal discussion of topics important to NATO allies (such as proliferation of WMD), but this needs to be balanced with the limited geographic overlap and the modest prospects for success it currently faces. The WEU has great interdependence and geographic overlap with NATO; however, NATO and WEU activities are not complementary. This, added to the low chance of success of the WEU dialogue, makes a NATO/WEU linkage positive in some respects but complicated in others.

The most effective linkage for NATO is with the EU's Barcelona process. The tasks of the two programs are complementary since some Arab countries have ruled out military cooperation in the EU framework for the foreseeable future, but not in NATO's. The prospects for NATO-EU synergy are improved by great interdependence and the high probability of success of the Barcelona process. While this linkage is by no means the only one possible, it is by far the most important for NATO. The WEU dialogue comes in a distant second in this regard.

However, any attempt to coordinate NATO and EU initiatives in the Mediterranean is likely to be marred by the lack of internal consensus within the EU over the development of a common foreign and security policy (CFSP). The EU summit in Amsterdam on June 16–17, 1997, underscored that the CFSP debate is still hampered by disagreement over defense and security policy in general and WEU-EU ties in particular.[22] The continued lack of internal agreement about the future of CFSP casts doubt on the ability of European states to reach an effective compromise on the EU's security and defense identity in the near future. This lack of consensus is bound to hamper the ability of the three institutions to work together across the policy spectrum, particularly on Mediterranean issues.

---

[22]Tom Buerke, "Europe's Leaders Clash on Defense Cooperation," *International Herald Tribune*, June 18, 1997.

Chapter Three

NATO'S MEDITERRANEAN INITIATIVE

Throughout the Cold War, NATO paid scant attention to developments in the Mediterranean that were not directly related to the Soviet threat. NATO began to focus on the Mediterranean in the 1960s with the establishment of the Expert Working Group on the Middle East and the Maghreb, and later—at a more political level—of the Ad Hoc Group on the Mediterranean. Composed of area specialists from allied countries, these groups conducted traditional monitoring of Soviet-related activities, as well as assessments of region-specific issues.

In the early 1990s, the Ad Hoc Group on the Mediterranean began to discuss the emergence of such new security risks in the Mediterranean as the proliferation of WMD and capabilities for their delivery, the growth of instability and extremism in North Africa, and the conflict in Bosnia. Concerned about the repercussions these risks could have on their security, Southern European allies began to call for a more robust NATO interest in the problems affecting the Mediterranean.

The lead in advocating an outreach program for the South was taken by Italy and Spain. These countries noted the rise of new regional security dilemmas and their possible effects on European and Alliance security. The Alliance's interest in the stability of the Southern Mediterranean and Middle East was included in NATO's Strategic Concept, which was adopted in Rome in November 1991. In large part because of pressure from Southern European countries, NATO decided at the December 1994 meeting of the North Atlantic Council "to establish contacts, on a case-by-case basis, between the Alliance and Mediterranean non-member countries with a view to contributing to the strengthening of regional stability."[1]

The purpose of the NATO Mediterranean Initiative is to achieve better mutual understanding with the countries to the South and to contribute to the strengthening of stability in the Mediterranean region, by making the Alliance's aims and objectives better understood.[2] Exchanges of views and information are designed to foster transparency and a better understanding of security issues of mutual interest. These aims are consistent with the NATO Strategic Concept, adopted in Rome in 1991, which states that the security policy of NATO is based on three mutually reinforcing elements: dialogue, cooperation, and the maintenance of a collective defense capability.[3]

On February 8, 1995, NATO announced that the Mediterranean Initiative would include Egypt, Israel, Mauritania, Morocco, and Tunisia. From the outset, NATO made clear it intended to adopt a "phased approach" and that new countries could be invited to join at a later date. In line with this, Jordan was invited to join later that year (the above-mentioned countries, including Jordan, are also known as the "dialogue" countries). Contacts between NATO and Mediterranean partner representatives began in May 1995. The first meetings were organized to explain the nature and purpose of the Alliance, to solicit the interests and concerns of the partners, and to schedule discussions on future steps.

Regular political discussions were later established to review topics of common interest, with agendas agreed upon in advance by NATO and the dialogue countries. Issues discussed in 1996 included political, social, and economic developments in the Mediterranean; peacekeeping; and opportunities for regional cooperation.[4] Each side may request additional meetings on a case-by-case basis. While NATO's Political Committee and North Atlantic Council determine

---

[1]*NAC Final Communiqué*, NATO M-NAC-2(94)116, issued at the Ministerial Meeting of the North Atlantic Council, NATO Headquarters, Brussels, December 1, 1994.

[2]*NAC Final Communiqué*, Ministerial Meeting of the North Atlantic Council in Noordwijk Aan Zee, May 30, 1995.

[3]"The Alliance's Strategic Concept," *1995 NATO Handbook*, pp. 40–42.

[4]Rodrigo de Rato, *In Pursuit of Euro-Mediterranean Security*, Working Group on the Southern Region, North Atlantic Assembly, May 1996, p. 10.

the Alliance stance on the dialogue, contacts between NATO and the dialogue countries have been exclusively managed by the International Staff on a bilateral basis, except in the field of information, for which NATO sponsors multilateral activities involving all of the dialogue countries.

Political discussions are accompanied by a set of measures to foster cooperation on more practical grounds. Cooperation is based on current NATO projects in the fields of scientific research, training in security and defense policy, and civil emergency planning (CEP). Dialogue partners are able, for instance, to participate in meetings organized by the NATO Science Committee as well as receive and disseminate information concerning the scientific activities of NATO. Peacekeeping courses offered by the Supreme Headquarters, Allied Powers, Europe (SHAPE) school in Oberammergau are also open to representatives of partner countries, while the NATO Defense College (NADEFCOL) in Rome is in the process of establishing contacts with corresponding institutions in dialogue countries to compare curricula and explore areas of cooperation. The NADEFCOL Deputy Commandant for Curriculum Planning visited college counterparts in Egypt, Tunisia, Israel, and Jordan in 1996. The International Staff regularly briefs dialogue countries on such topics as NATO's CEP. Involvement by dialogue countries in all cooperation activities is on a self-funding basis.

NATO's Office of Information and Press has also embarked on a public information campaign targeting opinion leaders of Mediterranean dialogue countries. Public relations are an important component of the outreach program, given the image problems NATO has in the region (the Alliance is still viewed with considerable suspicion in many circles in the Middle East and the Maghreb). Information-related activities have included seminars and conferences with the six dialogue countries as well as visits to NATO headquarters. While NATO has appropriated some funding for public information activities, the amount of money has been relatively modest and not sufficient to conduct a large outreach effort that could significantly affect perceptions in the dialogue countries.

INITIAL RESULTS AND PLANS FOR THE FUTURE

The first round of meetings, carried out over the spring and summer of 1995, has shown that the Alliance's negative image in the dialogue countries reflects a lack of information on NATO's goals, strategy, and operations. The Mediterranean Initiative's political discussions and the information campaign are aimed at eliminating such misperceptions and gaining the trust of dialogue partners. NATO has so far focused on bilateral discussions with ambassadors from partner countries in Brussels.[5]

The meetings to date have also made clear that some countries are more eager than others to engage in substantive discussion. Israel, Jordan, and Egypt have asserted their interest in exchanging views and cooperating on a wide range of issues, while the remaining countries have been more conservative in their approaches.[6] Arab countries remain reluctant to enter into substantive discussions on the proliferation of WMD, which in large part reflects unresolved issues within the Middle East peace process.[7]

After one year of regular discussions, the Alliance decided to evaluate the first stage of the Mediterranean Initiative and plan its further development. In late 1996 NATO produced a non-paper suggesting ways in which dialogue activities could be expanded. The non-paper was modest—it reaffirmed the intention to hold twice-yearly political discussions with the agenda to be jointly determined by dialogue countries and NATO. It also proposed multilateral "Med" briefings at the end of Alliance ministerial summits. The possibility of NATO presentations on crisis management and other topics of potential interest to dialogue partners has also been discussed.

A more significant development came at the NACC Madrid summit on July 8, 1997, when the NACC established the Mediterranean Co-

---

[5]Pedro Moya, *Frameworks for Cooperation in the Mediterranean Basin,* Subcommittee on the Mediterranean Basin, North Atlantic Assembly, October 1995, p. 42.

[6]Discussions with Israel, for instance, included terrorism, the spread of WMD and the development of military cooperation programs. See Moya, *Cooperation for Security in the Mediterranean: NATO and EU Contributions,* Subcommittee on the Mediterranean Basin, North Atlantic Assembly, May 1996, p. 9.

[7]This is another example of how the Middle East peace process affects NATO's Mediterranean Initiative. For a broader discussion of this point, see Chapter Five.

operation Group (MCG).[8] Tasked with the overall responsibility for the Mediterranean dialogue, the new committee will include representatives from the 16 allies. While the responsibilities and plans for the MCG structure remain sketchy, the meetings are expected to include internal discussions among allies, talks between allies and individual dialogue countries, and multilateral talks between allies and a number of dialogue countries.

The MCG will replace the Ad Hoc Group on the Mediterranean, since regional experts from member states will be allowed to participate in the meetings of the new committee. In practical terms, there is little difference between the Mediterranean Cooperation Group and the Political Committee—the chairperson of the Political Committee will also chair MCG meetings. The creation of the MCG is the clearest sign to date of NATO's willingness to raise, however modestly, the political profile of its Mediterranean Initiative.

Despite this slightly greater attention on the Mediterranean, the emphasis remains on gradualism, with the menu of offered activities to expand slowly and incrementally over time. Cooperation projects are likely to remain largely on a self-funding basis. More courses at NATO schools will be offered with particular emphasis on civil emergency planning and civil-military cooperation, along with medical evacuation workshops and civil protection seminars. Public information activities are expected to remain at their current levels over the course of 1997. NATO has also expressed an interest in involving dialogue partners in concrete military activities in the fields of peacekeeping and arms control and verification, although there is no clear consensus within the Alliance on what specific military activities ought to be opened to dialogue countries.

ATTITUDES OF MEMBER STATES

The political profile, depth of discussions, and scope of activities of the Mediterranean Initiative pale in comparison to the outreach package NATO devised for the East. The Mediterranean Initiative has been inhibited by internal squabbles over what specific projects

---

[8]*Madrid Declaration on Euro-Atlantic Security and Cooperation,* Heads of State and Government Meeting of the North Atlantic Council in Madrid, July 8, 1997.

should be included in the agenda, as well as by the reservations of the dialogue countries about rushing into cooperation with what have been traditionally perceived as hostile foreign militaries. However, the major constraining factor is the lack of a firm consensus within NATO on the long-term goals and utility of the Mediterranean Initiative.

The internal consensus currently supporting the discussions is a delicate one; member states differ on how the Mediterranean Initiative should evolve and on its importance and desirability. A day of talks on the Mediterranean dialogue with NATO country representatives yields differing perceptions of what the ultimate purpose of the initiative is. Some regard it as a simple public relations effort, others think of it as a useful channel to discuss security questions, while the more ambitious would like the dialogue to consider how NATO could address the security concerns of the Mediterranean partners.

Opposition to a more substantive dialogue has been expressed by a number of allies, especially those who are not directly exposed to security problems in the Mediterranean, who attach overwhelming importance to improving ties with the East, or who fear a conflict with other Mediterranean efforts (whether bilateral or multilateral) they care more about. Canada and Northern European countries, such as Norway, are not deeply involved in Mediterranean security issues. The United States has given only lukewarm support to the dialogue because of concerns that the dialogue might obstruct the Middle East peace process, while France has feared it might interfere with traditionally strong bilateral relations with North African countries. Paris also believes that the EU is better positioned to play the lead role in the Mediterranean. Several allies are also concerned that raising the dialogue's profile even further will elicit hostile reactions from the publics in the dialogue countries and from those Southern Mediterranean states that have been excluded.

The chief promoters of the Mediterranean Initiative are Italy and Spain; these countries feel that since the end of the Cold War, NATO has placed excessive emphasis on the East and neglected important security issues affecting the Alliance's Southern borders. In their original plan, Italy and Spain recommended the involvement of the International Staff in planning, the participation of the Political Committee in the discussions with Mediterranean partner countries, and the inclusion of Algeria in the dialogue. Italy and Spain (as well as Portugal) have also pushed for PfP-like cooperation—including joint military exercises.

Southern European pressure ultimately led to the creation of the dialogue; however, the original plans were considerably watered down after internal discussion. Substantive military cooperation, the inclusion of Algeria, and the direct involvement of the Political Committee in the talks have been excluded from NATO plans for the foreseeable future.

Discussions on the dialogue's membership were not contentious, with the exception of Spain's and Italy's support for Algeria, which was successfully opposed by other allies. Spain instead managed to include Mauritania in the roster of partner states, and the United States lobbied for the admission of Israel. There was more agreement on the exclusion of Syria and Libya because of their support for terrorism. The United States sponsored the entrance of Jordan into the dialogue in late 1995, managing to overcome the objections of other allies who wished to defer enlargement of the group to later stages.

One issue that has been the source of much controversy has been the "Partnership for the Mediterranean" (PfM) concept. During the Williamsburg ministerial in the Fall of 1995, then Italian Defense Minister Giandomenico Corcione proposed that NATO consider setting up a PfM. In late 1996, a senior Portuguese Foreign Ministry official also presented his country's vision of how the dialogue could eventually become a PfM.[9] Both the Italian and Portuguese PfM statements were received with considerable skepticism within the Alliance; several members fear that the PfM could result in a politically and economically costly endeavor that would divert resources from the East.

Although the enthusiasm expressed by Italy and Portugal for a PfM has recently waned with the lack of progress in the peace process, the

---

[9]See the paper by Ambassador Antonio Monteiro of Portugal, "NATO and the Mediterranean Security Dialogue (A PfP-Inspired Model)," presented at the Mediterranean Dialogue Seminar organized by the North Atlantic Assembly in Lisbon, December 5–6, 1996 (mimeograph).

idea of replicating the practical component of NATO's Eastern out-reach program is very appealing to Southern European countries for two main reasons. First, the political effect of establishing a PfM would be great, since it would imply that the strategic significance of the Mediterranean is as high as that of the East. The second and perhaps more important reason for Southern European support of the PfM is that such a program would give the Mediterranean Initiative a better chance of success. The political significance of labeling the program "PfM" is probably of less consequence than ensuring that NATO's expertise—in military cooperation for instance—is fully exploited. Military cooperation is viewed in Rome, Madrid, and Lisbon as a useful way of building long-term trust and minimizing harmful overlap between NATO's and other initiatives in the region.

Proponents of the PfM idea, however, will have to resolve the issue of how to deal with different degrees of interest in military cooperation among the dialogue countries. If Mediterranean partners were allowed to freely choose the extent of their involvement with NATO as PfP members do, Israel would probably request more collaboration than any Arab partner. Self-selection could create an asymmetry in treatment, with Israel being the most eager to expand cooperation. This would contradict the objective of improving NATO's relations with the Arab world. While this is less of a concern as long as the initiative focuses on political discussions, it could be problematic if the agenda were extended to include military cooperation.

U.S. ATTITUDES

The success of the NATO Mediterranean Initiative also depends on the regional policies of its most influential member states, first and foremost the United States. Thus far, the United States has adopted a reserved stance toward the initiative.

American reluctance to contemplate a major shift in the initiative is related to the fact that, currently, the most important item on NATO's agenda for the United States is enlargement. Washington does not want NATO to have its attention distracted from eastward enlargement and relations with Russia. The United States also fears that the initiative may interfere with the Middle East peace process. U.S. opposition to the participation of dialogue partners in NATO environmental projects reflects Washington's concerns that these activities would divert attention from the environmental component of the peace process.

At the same time, it is in the interest of the United States to counter the rise of nontraditional threats—such as the proliferation of WMD—both in the Mediterranean and globally. The NATO dialogue is a useful way both to promote internal discussion on such topics and to effectively communicate any Alliance decisions on these topics to Southern partners. Joseph Kruzel, former U.S. Deputy Assistant Secretary of Defense for European and NATO Affairs, proposed in 1995 that NATO should consider strengthening the dialogue by including such PfP-style activities as multinational military exercises and military training.[10]

U.S. attitudes will critically affect the initiative's future—U.S. opposition to the CSCM and the subsequent demise of this project is a reminder of such influence. During President Clinton's first term, former U.S. Secretary of Defense William Perry frequently called attention to the growing importance of security problems in the Mediterranean. Since his departure, U.S. interest in the Mediterranean has seemed to wane, in part because of Washington's preoccupation with enlargement. However, any intensification and expansion of the Mediterranean dialogue will need strong American backing if it is to have a reasonable chance of success.

PROSPECTS

The intra-Alliance bargaining process has produced an initiative that is based on an incremental, case-by-case approach. Including more countries as partners and expanding to more concrete cooperation have not been ruled out in principle, but it is generally accepted that any intensification of the dialogue should occur gradually.

The dialogue could be more useful both to NATO and to its Mediterranean partners. It could, for instance, become more relevant to NATO's goals of projecting regional stability and crisis management, and make better use of Alliance expertise in certain key areas, such as

---

[10]See Joseph Kruzel, "The New Mediterranean Security Environment," presentation at the Institute of National Strategic Studies (INSS)/AFSOUTH Conference on Mediterranean Security, Naples, February 27–March 1, 1995 (mimeograph).

peacekeeping and civil emergency planning. Since counterproliferation is one of the new missions of the Alliance (as laid out by the
Policy Framework of 1994), the dialogue should, in principle, include
this topic in political discussions with partners. The participation of
Egypt, Jordan, and Morocco in Implementation Force/Stabilization
Force (IFOR/SFOR) and the support the mission generated among
the publics of these countries also suggest that peacekeeping may be
an area in which NATO can usefully expand cooperation with the
dialogue countries.

Even if the Alliance decided that the dialogue should change in some
form, it is not clear exactly how this should occur. When crafting a
viable strategy for the dialogue, several constraints need to be taken
into account. Discussions with NATO officials and with observers of
the Mediterranean dialogue have pointed to three operational obstacles in the way of a more substantive outreach program: (1) financial and organizational limitations, (2) opposition to extending
the dialogue to Algeria, and (3) opposition to military cooperation.
These are directly or indirectly related to a lack of agreement within
the Alliance on what the dialogue's agenda should include. In turn,
such lack of agreement is an obvious symptom of the absence of a
clear, well-articulated vision of NATO's role in the Mediterranean—a
role that is shared by all member states. In addition, different ways
of dealing with these constraints are not equally effective. The key to
successful change lies in choosing a course of action that will make
the constraints less binding while at the same time allow the intro-
duction of substantive items to the dialogue's agenda.

NATO also needs to consider external factors when assessing the effectiveness of the dialogue: A number of other initiatives exist in the
Mediterranean, and the effectiveness of NATO's initiative will depend to a large extent on how well it fits into the broader framework
of regional cooperation. While NATO has asserted the importance of
making institutions "interlocking," so far it has done little to promote
such a development regarding the Mediterranean. The question of
coordination with other institutions is likely to become more important over time. The pitfalls of duplication and competition will intensify as other frameworks for Mediterranean cooperation become
more substantive or expand their scope.

Coordination with the EU's Barcelona process and other initiatives would be easier if NATO articulated a coherent long-term strategy for its outreach program. Strategic coordination—a process to align the long-term goals of the various initiatives—is the most useful mechanism to ensure the viability of NATO's outreach program in the long run. By definition, this type of coordination would be feasible only if NATO developed its own strategic vision for the Mediterranean.

While NATO has been focusing on the need to correct misperceptions within Arab countries, little has been done to address the information gap that exists between NATO and other European security and political institutions involved in the Mediterranean. Partly as a result, some officials working in European institutions view the NATO initiative as a tool of U.S. foreign policy, aimed at undermining Europe's efforts to establish a presence in the region. Occasional and informal exchanges of ideas between NATO and the EU Commission leadership have been viewed with suspicion by the European Council. This bias, compounded with bureaucratic priorities and jealously guarded autonomy on the part of both NATO and European bodies, hampers the establishment of mutual trust—a precondition for coordination between institutions. The image of the NATO initiative as a ploy of a competitive U.S. foreign policy is inaccurate; the main proponents of NATO involvement in the Mediterranean are Southern European countries—the same ones pushing for a greater EU role in the region. These biases and misperceptions need to be corrected through more extensive information exchanges between NATO and other key actors in the Mediterranean, particularly the EU.

Achieving internal consensus on the Mediterranean Initiative's long-term purpose is not only necessary to ensure coordination with other frameworks, but also critically affects the dialogue's own chances of success over the long run. A clear goal for the dialogue can be articulated only when important differences that exist within the Alliance are resolved. A qualitative improvement in the initiative requires strategic thinking, and strategic thinking implies moving away from the incremental approach that appeals to skeptical countries. The key task for Southern European members is to achieve broad agreement within the Alliance for significantly expanding the scope and purpose of the initiative.

The chances that such consensus will be reached depend crucially on the position of the United States. If the United States placed its political weight behind efforts to substantially upgrade the initiative, support for it within the Alliance would greatly increase. The lukewarm approach adopted so far by the United States toward the initiative needs to be transformed into much more active political support and backing by Washington. Otherwise it will be difficult to seriously develop the initiative.

At the same time, care needs to be taken to minimize the dialogue's interference with the Middle East peace process. In fact, the effects of the Middle East peace process have to be taken into account by NATO planners when considering how to upgrade the initiative. Judging from the direct repercussions of the Middle East peace process on the fortunes of other Mediterranean frameworks, particularly the Barcelona process, there is a trade-off between increasing cooperation with dialogue countries and exposing the initiative to the controversy of unresolved regional disputes.

There are also operational questions that need to be addressed about the initiative's short-run strategy. Decisions on whether to include more-concrete military cooperation, to expand the dialogue to Algeria, or to provide some financial support for partner participation will significantly influence the course of the initiative in the near term. Increased funding may prove necessary to encourage a satisfactory response from dialogue partners to the cooperation programs devised by NATO. Financial constraints play a larger role in the decision to cooperate with NATO in the case of Southern Mediterranean countries than they do for PfP countries, yet the latter have access to NATO funding while the former generally do not. Unless this issue is resolved, the lack of some kind of financial backing could undermine the credibility of the dialogue.

The Alliance also needs to recognize that expanding and intensifying the dialogue is likely to bring a set of new complicated issues, such as the self-selection problem in the case of a PfM. However, all these considerations stand a better chance of receiving systematic and thorough attention if NATO devises a comprehensive Alliance strategy for the South, of which the initiative would be a significant, but not exclusive, part.

Chapter Four

PERSPECTIVES OF THE DIALOGUE COUNTRIES

The success of NATO's Mediterranean Initiative will to a consider-able extent be determined by the degree to which it addresses the concerns and fears of the dialogue partners themselves. How do they regard the initiative? To what extent does it address their secu-rity concerns? How can the initiative be made more relevant and ef-fective?

As a general principle, none of the dialogue states is opposed to the initiative per se, and each of them regards it with a varying degree of interest. At the same time, there are a variety of uncertainties, ques-tions, and concerns that need to be addressed if the dialogue is to achieve its objectives and promote greater transparency for and un-derstanding of NATO's goals.

One of the critical issues for the dialogue countries is the composi-tion of the group itself—why each was chosen to participate in the dialogue and why others were excluded. Each of the six countries realizes that it was chosen largely because it is perceived to be a moderate, Western-looking, constructive (as defined by the West) participant in regional affairs. Furthermore, all six have diplomatic and political ties with one another, which is no small matter given the fractious quality of political life in the Middle East and North Africa. The states that have been excluded from the initiative are significant: Algeria, Lebanon, Libya, and Syria. Their exclusion is understandable from a Western point of view, yet it is a problem nonetheless. For on a general conceptual level, the exclusion of any Arab state is considered unacceptable by other Arab states in any set-ting where multiple Arab states are represented. This absence is further magnified by the inclusion of Israel in the initiative. Any willingness by Arab states to participate in meetings where Israel is represented and other Arab states are not is seen by many Arab states as a breach in Arab solidarity.

Put simply, despite Iraqi aggression against Kuwait, or Syria's preference for Iran instead of fellow Arab state Iraq in the Iran-Iraq War, Arab states still "stick together" or would have outsiders, as well as other Arabs themselves, believe this to be the case. This dynamic is a complicated one, although a gap between political ideals and practices is not unique to the Arab world. Nevertheless, each Arab state realizes that as a temporary measure at least, it is possible to participate in a forum with Israel from which other Arab states have been consciously excluded. However, there are limits to the willingness of most Arab states to be involved in such activities. At this point, this is not a major problem because NATO's Mediterranean Initiative is so little known in the Arab world. However, should the dialogue be expanded and deepened, this issue could become a more serious impediment to greater cooperation.

For the moment, all participant countries understand that they must have working relations with one another, that they be commonly accepted by all of NATO's members, and that they must share certain broad political perspectives on the region. Thus, temporarily at least, the exclusion of Algeria, Lebanon, Libya, and Syria is tolerated, although never fully ignored or sanctioned, by Arab dialogue participants. Syria and Israel are still in conflict, and Syria still refuses to participate in American-brokered peace efforts in the region. Lebanon is today largely dominated by Syria and closely toes the Syrian line. Algeria is embroiled in a de facto civil war and Libya remains an explicit outcast from the West and an implicit one from the Arab world. Thus, even if these states were willing to participate, there would be significant opposition to their involvement by important NATO member states.

Each of the six dialogue countries participating in the Mediterranean Initiative is well aware of NATO's enlargement to the East. Expected to cost billions of dollars, NATO's Eastern expansion stands in stark contrast to its Mediterranean Initiative, for which few funds have been committed. Despite obvious differences between the two efforts to promote stability, the predominant impression in the minds of the dialogue countries created by the disparity in resources devoted to the two efforts is that the North is "worth" more than the South. Or put differently, that Europe is accorded greater importance by NATO and its member states than is North Africa or the Middle East.

Given that the primary movers for the Mediterranean Initiative within NATO are Italy and Spain, the Arab states of North Africa and the Middle East perceive the initiative as representing a policy of "ambivalent engagement." Many see that the initiative is primarily designed to prevent unfettered migration, drug smuggling, terrorism, and other "unhealthy" influences from crossing into Europe from North Africa. While this view relates more to the policies of the EU than to those of NATO, it also carries over to perceptions of the purposes of NATO's dialogue as well. Many dialogue countries feel that NATO is essentially seeking a way to keep the North African states and their problems at arm's length, rather than genuinely trying to engage in closer partnerships with them.

Not all NATO member states regard the initiative with the same degree of enthusiasm, which is another concern. The greatest support for the dialogue comes from NATO's Southern members. But the Northern members have far less interest in the initiative. Furthermore, not all of the Southern NATO countries support the initiative with equal ardor. France's position, for instance, differs from that of Italy and its commitment is tempered by its own idiosyncratic perception of its interests in the region as well as its attitude toward NATO's transformation and evolution.

NATO's negative image in the Middle East and North Africa also presents problems. For many years, NATO was regarded as being emblematic of, if not virtually synonymous with, the cultural and political depredations of the West. NATO was seen as the West's main military instrument against Soviet power. This notion of "being Western" is at odds, to some degree, with the characterization that the Mediterranean dialogue countries have of themselves (Israel excepted). With the exception of Jordan, the dialogue countries are uniformly Arab, Islamic, African, and, in all cases, Israel included, the victims of Western colonialism. Despite the fact that the heads of these states are in some cases heavily oriented toward the West (e.g., Morocco's King Hassan and Jordan's King Hussein in particular), these leaders have taken pains to emphasize their own countries' unique religious and cultural heritages. King Hassan, for example, repeatedly makes reference to the fact that he is a direct descendant of the Prophet Mohammed. Thus, at the symbolic level, the initiative comes from the very "cockpit" or headquarters of the West and is directed at a number of countries to the South whose ties with the West have historically been characterized by distrust, conflict, and betrayal.

This issue reflects a great sensitivity throughout the Arab world to initiatives from or formalized ties with the West. Middle Eastern states are highly sensitive to what they regard as Western feelings of religious, cultural, and political superiority toward the Middle East. There is great awareness throughout the region of such historical landmarks as the Crusades, the support of Western-style Zionism over Arab-oriented Palestinian nationalism, and the continued support by the West for corrupt, despotic, and unpopular regimes throughout the region. These asymmetrical power relationships in which French democracy could turn its back on elections in Algeria, or the democratic United States could go to war to "liberate" Kuwait for the Emir, highlight historical ties that have rarely favored the peoples of the Middle East. Former NATO Secretary General Willy Claes' unfortunate statement in February 1995 that Islamic fundamentalism had emerged as perhaps the greatest threat to Western security since the collapse of communism in Eastern Europe has tended to reinforce such impressions.[1] To compare a mode of spirituality with a credo premised on atheism was deeply offensive and insulting to Muslims everywhere, although hardly surprising to them. Despite constant denials from the West, many Middle Easterners remain convinced that Claes merely articulated what most in the West feel anyway and that the West has never really come to terms with the Islamic world or the Middle East in the first place. They believe that the West sees their region only as a place to be exploited, defended against, or ignored. Although this does not by any means doom NATO's Mediterranean Initiative, it does highlight the challenge confronting NATO as it attempts to create partnerships with

---

[1]See "NATO Chief under Fire for Islamic Remark," *International Herald Tribune*, February 15, 1995.

those whom some of its member states have exploited far more systematically than they have helped.

To confirm this, one has only to realize how attentive Middle Easterners are to the utterances of French politicians, such as to statements from Jean-Marie Le Pen, or even to Samuel Huntington's widely read writings on the "clash of civilizations."[2] Although the latter is a dispassionate attempt by an important American thinker to understand the changing contours of international politics, it has been seen by many in the Middle East as being racist and provocative.

This raises another issue of concern to many of the dialogue countries—what is NATO really seeking in this security initiative? In the Middle East, NATO is known primarily as a bulwark against Soviet expansionism. But the Soviet Union has disappeared from the scene and the dialogue countries of the South present no security challenges to the states of the North. At the same time, the states of the South are not confronted with the sort of external security threats that NATO could help them resolve.[3] Thus, NATO's attempt to initiate a security dialogue when there are no evident, high-level external security problems that need to be resolved strikes many Middle Easterners as strange, if not confusing. Although it is quite reasonable for NATO to seek to promote a relationship with the Southern Mediterranean countries based on common security considerations, these considerations must be identified and catalogued from the outset. Indeed, to some in the Middle East, NATO and the West *are* the security threat. Thus, continued deep-seated suspicion about the dialogue is likely to persist until its goals can be better articulated and more broadly understood.

Realistic conceptualizations of security must be at the heart of a successful NATO dialogue. Yet none of the dialogue countries face significant external security threats. Whereas NATO's view of security

---

[2]See Samuel Huntington, "The Clash of Civilizations," *Foreign Affairs*, Summer 1993, pp. 22–49.

[3]This point is recognized in a commentary entitled "No Mediterranean PfP," *Defense News*, October 21–27, 1996, p. 32. The prior week there was an article evaluating the virtues of PfP-type military relations entitled "NATO Extends Hand to Middle East Nations," Ibid., October 14–20, 1996, p. 1. Both pieces recognize the lack of specificity characterizing the NATO initiative.

was forged in the crucible of a bipolar world characterized by political, military, and economic rivalry between the two antagonistic blocs, there is no comparable external threat hanging over any of the dialogue countries. This is not meant to suggest that the region is free of conflict—for clearly it is not. Nevertheless, security to the dialogue countries primarily reflects internally inspired subversion and politically motivated violence, terrorism, or revolution, not external aspirations for regional hegemony that could successfully be thwarted through a partnership with NATO. The external predations of Libya are regarded as being somewhat trivial in comparison with the internal disintegration currently afflicting Algeria.

Another important consideration is the role of the military in virtually all of the dialogue countries, Israel excepted. In each dialogue country, the civilian sector rules only because it has been able to effectively control the military while being able to count upon its support when necessary. Such control in nondemocratic polities requires exceedingly deft and careful management by a head of state. He ignores or takes the military for granted only at his own peril. Thus, before any type of deep-seated security arrangements can be crafted with these states, NATO must ensure that such an effort will not unwittingly weaken the ruling elite. In these countries, the support of the military is an essential prerequisite to remaining in power. On the one hand, the head of state cannot live with the military (cannot share power too *broadly*). On the other hand, he cannot maintain power without the support of the military. Any security dialogue with these countries must be sensitive to these dynamics.

Another issue that cannot be ignored is the radically different political systems that exist in NATO member states as opposed to the dialogue countries to the South. Although some of these countries have experimented with limited forms of democracy with occasional modest success—most notably Jordan—with the exception of Israel, none of these states is a democracy as the term is widely understood in the West. This raises important questions that NATO will eventually need to face. Should NATO promote democratization in these countries? To what degree should the legitimacy of political elites be a concern of NATO? These and other exceedingly difficult questions must be addressed if the dialogue is to move beyond the ceremonial level.

Finally, the NATO dialogue is hindered at the outset by the fact that it is the most recent initiative in a long list of Euro-Mediterranean initiatives that have generally led nowhere. This point was emphasized in an interview with Tunisia's highly respected Minister of Defense Habib Ben Yahya (then Minister of Foreign Affairs), who catalogued a prodigious list of such initiatives over the past 30 years.[4] The vast majority of these dialogues, he noted, had been initiated by the Europeans and most of them never accomplished anything. Thus, if a certain measure of skepticism or ennui is evident among Middle Easterners about NATO's Mediterranean Initiative, this skepticism must be viewed against the background of the meager results from the multitude of Euro-Mediterranean initiatives that have preceded it.

PERSPECTIVES OF INDIVIDUAL DIALOGUE COUNTRIES ON THE INITIATIVE

The dialogue countries are not a monolithic group. Each has its own interests, concerns, histories, and aspirations. Thus, these countries must be considered not only as individual regional actors, but also as part of a broader and even more complex regional fabric in which they all relate to one another. In addition, the architects of the NATO initiative confront significant policy challenges because different Middle Eastern states have different ties with assorted NATO members and their expectations of the dialogue vary.

Egypt

The Egyptian approach to the initiative differs from that of other Arab states. The reason for this is that Egypt is acutely aware of its status as the largest and most powerful state in the Arab world. Although Egypt was the first Arab state to make peace with Israel, it has also been one of the most outspokenly critical of Israeli policies. Egypt has always felt a certain obligation to criticize Israel on behalf of the Arab world in a fashion not meant to mitigate its peace with Jerusalem, but rather to use its peace with Israel as a means to be more influential in affecting Israeli regional behavior. Although this

---

[4]Personal interview, Tunis, October 1996.

policy has not been very effective, it is on this basis that Egypt and Israel will participate in the NATO dialogue.[5]

There are no meaningful security challenges that NATO can help Egypt resolve. Nonetheless, just as is the case with all of the other Mediterranean frameworks, Egypt is quite willing to continue its discussions with NATO as a way to improve ties with the member states of NATO. NATO represents not only a strategic organization, but also a collection of wealthy and powerful Western states, many of which are proximate to Egypt. Thus, Egyptian involvement in the dialogue is perceived by Egypt, just as it is by the others, as an opportunity to improve its ties, as well as its economic prospects, with these countries. Understanding this economic motivation is vital to grasping why countries of the Middle East and North Africa want to participate in the dialogue. While NATO is talking about security and promoting better ties with its neighbors, the dialogue countries are less concerned about security and more concerned about improving ties with wealthy neighboring states who happen to be NATO members. This key issue cannot be overstated, because all of the Southern dialogue countries aspire to greater economic cooperation with and closer ties to Europe. Thus, involvement in the NATO dialogue should be seen as one element in a broader effort by the dialogue countries to establish closer ties to Europe.

Despite this, there are few economic benefits from participation in the initiative. Nonetheless, Egyptian participation and involvement are highly symbolic, given Egypt's unique role in the Arab world. Egypt will have no qualms about supporting low-level military cooperation in such areas as rescue attempts, drug interdiction, fighting terrorism, and so forth. This could happen either through NATO or bilaterally and regionally, given Egyptian ties with the states proximate to it. The symbolism of involving Israel in such activities is quite important for Egypt. While Jordan, Morocco, Tunisia, and Mauritania might be more willing to collaborate with Israel under the right circumstances, it is hard to envision any circumstances under which Egypt would be willing to entertain such collaboration

---

[5]For a thoughtful and highly critical discussion of the NATO initiative, see "Enlargement of Western Security Institutions: A Non-Western Perception," by Hani Khallaf, Deputy Assistant Foreign Minister for Security Organizations in Europe (unpublished).

before a complete resolution of the Arab-Israeli dispute in a fashion that would not only satisfy Palestinian nationalists' aspirations, but also would include such outlying Arab states as Syria, Lebanon, and even Libya as part of a final agreement. This commitment speaks to Egypt's perceived role as the leader of the Arab world, and although other Arab states may even resent attempts by Egypt to dominate discussion in multinational, Arab forums, Egypt's commitment to these issues and its conviction that it is the spokesperson for the entire Arab world is widely regarded as important and immutable. Thus, in responding to the substance of the NATO initiative, Egypt is particularly critical of the exclusion of other Arab states, such as Libya and Algeria, sensitive to the subordinate status of Arab states in such Euro-Mediterranean discussions, critical of the emphasis on security concerns rather than economic and cultural issues, and is unhappy with the tendency to use terms like "partnership" when in fact the Europeans, in Egypt's view, have no intentions to build a true partnership but rather wish to "protect" themselves from migrations by their Arab neighbors.

Israel

In general, Israel is quite pleased to be included in the initiative. Indeed, it is often forgotten that Israel for many years sought membership in NATO and was repeatedly rebuffed. Although Israel's strongest external supporter is widely recognized to be the United States, during the state's early years and throughout much of the 1950s, Israel relied far more on Europe for military support than on the United States. In 1956, Israel even collaborated with Britain and France in a brief war against Egypt, and Israeli military ties with France in particular were far deeper than is often recognized. One by-product of the Arab-Israeli peace effort has been the gradual acceptance of Israel within the Arab world, as well as movement toward Israel's broader regional integration. Thus, inclusion of Israel in the initiative was a logical next step in which the region began to be seen as a strategic entity, and not merely as a collection of idiosyncratic states in conflict with one another.

NATO's Mediterranean Initiative was launched at an important and propitious moment in Middle East history, since Arab-Israeli peace efforts permitted Israel and several erstwhile Arab foes to participate in this activity together. However, Israeli security policy has evolved dramatically since the 1950s and 1960s when Israel vainly sought acceptance by NATO and Europe. Israel has forged a variety of deep and abiding strategic, political, economic, and security relations with important countries as diverse as China, India, the United States, South Africa, and others. Thus, while Israel finds the idea of a dialogue with NATO appealing, it does not actually need such a dialogue and, for the moment at least, appears to perceive it more as a potentially interesting diplomatic opportunity than as a strategic necessity.

Israel's view of the dialogue is upbeat and positive, but hardly enthusiastic. It will make few sacrifices to enhance the dialogue, will risk nothing, and will use few resources on its behalf. Nonetheless, Jerusalem will remain open to good ideas. Israel has a number of strategic assets and capabilities that it would be willing to commit to a more formal relationship with NATO, but it is unlikely to work terribly hard to make this dialogue a higher priority than it currently is. Finally, Israel is coming to terms with being part of the Middle East. Thus, Israel is quite happy to be part of the initiative because the other five dialogue countries are Arab and the symbolism of Israel's regional acceptance and integration is appealing. As a consequence, Israel will certainly support any dimension of the dialogue as long as it does not interfere with Israel's broader regional or extra-regional ties or relations.

Jordan

Jordan, under the leadership of King Hussein, has always had to compensate for its vulnerability, which results from a weak economy, irrevocable and hard-to-control ties to the Palestine issue, and proximity to a variety of rich, powerful, and at times, unstable neighbors (e.g., Iraq). At the same time, Jordan has always sought improved ties with the West as a central and important feature of its foreign and national policy. King Hussein's Western orientation is clear. Thus the NATO initiative has provided Jordan with yet another opportunity to improve and deepen its ties with the West. Although the fact that Jordan is not proximate to the Mediterranean can be dismissed as being irrelevant to the initiative, questions can be raised about the nature of challenges to Jordan's security and whether

NATO can help to address these challenges. Nonetheless, Jordan clearly regards the initiative in highly favorable terms, although what it actually expects from the initiative remains unclear.[6]

Certainly on a low level, Jordanian involvement may well permit closer ties between Amman and a variety of European states. Here there is little risk of a conflict between multilateral ties mediated through an international organization and bilateral relations—because nothing that emerges from this dialogue is likely to be of great concern to any NATO member, all of whose bilateral ties with Jordan are unlikely to be disrupted. Jordan will assuredly participate in the dialogue, as well as support continued Israeli involvement for as long as Israeli policies, in a general sense, do not interfere with Arab ties with the Jewish state. However, if Israeli Prime Minister Benjamin Netanyahu continues to accede to the demands of the extreme Israeli right, and forges policies that crudely and unambiguously undermine prior agreements with the Arab world, then King Hussein will undoubtedly choose to join ranks with other Arab states to express his displeasure with Israel. King Hussein is unlikely to be the first to reach this conclusion, nor would he be the last to criticize Israeli policies. Nonetheless, in terms of the initiative with NATO, it is clear that Jordanian involvement will continue as long as Jordan perceives a benefit to itself from such involvement, even though its expectations are not likely to be very high.

Mauritania

Mauritania is by far the most remote and weakest of the dialogue countries. The leadership in Nouakchott is very pleased to be included in the dialogue, and although its capabilities are limited, Mauritania's role in the dialogue should not be taken for granted. As do the other members of the dialogue, Mauritania has a peaceful relationship with Israel. Indeed, it is noteworthy that during the Hebron period, Mauritania was one of the few Arab states that did not allow its opposition to Israeli policies to interfere with or disrupt its relationships with Israel. In this period, Mauritania opened an

---

[6]For a January 1997 discussion of Jordan's perspective on the NATO dialogue, see "Security and Cooperation: Jordan's Position," available from the Jordanian Embassy to Belgium/NATO (unpublished).

interest section in Tel Aviv and continued on a course of business as usual with Jerusalem.

As is the case with other members of the dialogue, Mauritania has little interest in a traditional security relationship with NATO. Rather, its main reason for engaging in this dialogue is to enhance its relationship with the West and gain greater Western economic assistance. The dynamics here are not unlike the dynamics for the other countries. However, given Mauritania's geographical location, its needs are somewhat different from those of its neighbors. Some Mauritanians are concerned about a threat from Morocco. This threat has its origins not only in the Western Sahara conflict, but also in a North-South relationship with Morocco, which is a large and powerful state from the perspective of the smaller and far less well-developed Mauritania. Whether the NATO initiative can help to re-assure Mauritania is questionable. However, greater interaction within the framework of NATO's Mediterranean Initiative could possibly help to allay some of these fears—which would in itself be an important contribution to regional security.

In this struggle for national development, Mauritania has important policy challenges. For example, it has attempted to democratize and has held elections that did not receive wide publicity, but that were generally considered fair and impartial. Mauritania also struggles under the legacy of having, until quite recently, sanctioned slavery. Although the government feels that it has taken more than adequate measures to abolish slavery, there are those outside of the country who accuse Mauritania of continuing to maintain this tradition in a sporadic and unsystematic fashion. Unfortunately, from the perspective of the Mauritanians, those who are critical of Mauritania seem far more determined and vocal than those who wish to help it. Thus, Mauritania probably hopes that the dialogue with NATO will promote greater familiarity with the country and help it obtain greater financial assistance to pursue its national development in a more successful fashion.

Please keep in mind that Mauritania is bordered on its south/southwest by Senegal. It is part of a belt of states that include Mali, Niger, Chad, and Somalia, in which the Arab world gradually is transformed into black Africa. Thus, Mauritania has a variety of other political involvements and commitments to its south that may be irrelevant in its dealings with the North, but that do present it with political challenges unlike those of the other dialogue partners.

Morocco

Morocco is generally willing to engage in any type of political, diplomatic, or economic activity with its powerful neighbors to the North to enhance its relations with Europe. Indeed, Morocco's King Hassan has been most effective in emphasizing different aspects of his own national priorities depending upon the interlocutors with whom he is dealing. Thus Morocco is not only an Islamic, African, and Arab state, but also one regarded by the European powers as reliable and cognizant of its own needs. However, Morocco creates a number of problems for its neighbors to the north as a result of its own developmental problems. The most important is the significant outward migration of Moroccans seeking better economic opportunities in the wealthier states of Southern Europe.

As with the other dialogue countries, Morocco regards the initiative as an opportunity to improve its relations with a group of powerful Western states. Morocco does not have any perceptible external security needs that NATO could help it resolve. Morocco is far less concerned with an external threat from, say, Libya than it is with the repercussions of internal instability in, for example, Algeria. Its most pressing security concern is the potential emergence of an indigenous anti-regime political movement based on Islam that seeks to overthrow the current Moroccan government and replace it with some sort of Islamic state. This type of internal threat cannot be effectively suppressed militarily. Most of the dialogue countries believe that the key to internal security—i.e., the antidote to Algeria-type chaos—lies in improved domestic development.[7] It is absolutely essential for the NATO states to appreciate this, because when Morocco thinks of security it thinks of an internal threat posed by an increasingly restive Moroccan population, rather than an external threat to its national borders. In short, there is an inherent contradiction between what NATO offers and what Morocco feels it needs.

---

[7]See Simon Serfaty, "Algeria Unhinged: What Next? Who Cares? Who Leads?" *Survival*, Volume 38, Number 4, Winter 1996–97, pp. 137–153.

It is reasonable to expect that Morocco will continue to participate in discussions with NATO as long as NATO is interested in maintaining these discussions. Nonetheless, it is equally clear that what Morocco and other Arab dialogue states would like most is something that NATO is not in a position to deliver: external economic and political support designed to help them deal with internal challenges. For Morocco, national security revolves first and foremost around an ability to satisfy the needs of the Moroccan people, which include housing, transportation, nutrition, health, electrification, civil and criminal justice. These are needs that NATO is not particularly well-suited to address.

Tunisia

As is the case with the other Francophone North African states Morocco and Mauritania, Tunisia regards the initiative from a somewhat different perspective than the states in the Eastern Mediterranean. The North African states, particularly Tunisia and Morocco, are virtual suburbs of Europe. Tunisia is closer to Italy than it is to Egypt, and thus it is acutely aware of its dependence on Europe. At the same time, Tunisia is situated in a particularly difficult neighborhood that includes Libya and Algeria, and thus it is no less sensitive to its status and role as an Arab, African, and Islamic state. It is these cross-pressures that have made Tunisian foreign policy so precise and carefully designed. Tunisia is quite happy to participate in any European-sponsored dialogue. Although NATO is certainly not a European organization exclusively, the main impetus for the initiative comes from Southern European states. Thus, Tunisian involvement in the initiative is predated by countless instances of participation in Euro-Mediterranean dialogues, discussions, study groups, commissions, reports, and all of the other accoutrements of dialogue and diplomacy by neighboring states to the North, whose history has been more one of exploitation than one of providing genuine assistance to North African neighbors.

Despite its realism, Tunisia is still the leading dialogue country in North Africa. Its levels of development are the highest in the region, while its ability to negotiate beneficial agreements with Europe is the most well-developed. Tunisia has agreed to participate in the dialogue in a skeptical but, nonetheless, collaborative and open fashion.

Tunisia was the first state in North Africa to sign a formal trade agreement with the European Union and thus its perception of the NATO initiative is less as a security-type arrangement and more as an agreement that is part of a more complex set of Tunisian-European understandings.[8]   While Western officials sometimes think that Tunisia requires or wants some sort of security guarantees from Europe, given its proximity to Libya and Algeria, nothing could be farther from the truth.   Indeed, it is testimony to Tunisian diplomatic sophistication that Tunis has been able to maintain tolerable working relations with all of its neighbors.   Tunisia may be concerned about its neighbors, but it is not necessarily afraid of them.   Thus, NATO should not be deluded into thinking that external security is first and foremost on Tunisia's mind.

MIDDLE EAST POLITICS AND NATO'S INITIATIVE

A key factor contributing to the acceptance of NATO's Mediterranean Initiative was that, after many years of conflict, both active and passive, a variety of Arab states littoral to the Mediterranean region had finally made peace with Israel.   The level of ties, the origins of the peace between these states and Israel, and the character of their current relations vary significantly.

For example, while Egypt was the first to formally conclude a peace treaty with Israel, Egyptian-Israeli ties have never been terribly positive, and Israeli-Egyptian relations are best characterized as being correct, if troubled and tense.   This differs dramatically from Israeli-Jordanian relations, which were exceedingly close from the moment peace was concluded.

For many years, before the formalization of peace between Israel and Jordan, the Jordanians maintained covert and constructive ties with Israel.   Once a full-fledged peace agreement blossomed, Israeli-Jordanian ties quickly became closer than those enjoyed by Israel with any Arab state.   Cooperation on a number of levels, including military, developed with remarkable speed after the conclusion of the peace, and from the Israeli perspective, Israeli-Jordanian

---

[8]See "Tunisia Signs Trade Agreement with European Union," *Tunisia Digest*, Embassy of Tunisia, Washington, D.C., October–December 1996, p. 5.

relations may serve as a model for ties that Israel would like to have with other Arab states.   Unfortunately, as discussed below, other Arab-Israeli ties have deteriorated, as have broader Arab-Israeli relations as a result of the stagnation of the peace process.

Ties between Israel and Morocco are also close.   Morocco's King Hassan quite prudently conducted a policy based on realism and pragmatism so that his ties with Israel predate by many years the conclusion of a formal peace agreement.   Thus, when peace between Israel and Morocco finally was codified, the world came to see a public face of a relationship that had gone on privately for many years.   Israeli-Moroccan relations have been both enduring and strong, as well as absent of any conflict other than muted and symbolic, Pan-Arab type rhetoric about Israel from Rabat on rare occasions.

To a lesser degree, this relationship typified Israeli-Tunisian relations as well.   Tunisia is well known for having a constructive and rational foreign policy that has shrewdly been designed to minimize Tunisia's vulnerability to the changing political vicissitudes and fortunes around it.   Sandwiched as it is between Libya and Algeria and closer to Sicily than to Egypt, Tunisia has always been vulnerable to more powerful neighbors and thus has always relied on pragmatic and carefully developed foreign policy orientations.   This was the case in its dealings with Israel as well.   For some time before the formalization of a peaceful relationship, Tunisia and Israel enjoyed reasonable ties with one another.   Although Tunis had been the extraterritorial headquarters of Yasser Arafat and the PLO, this in no way indicated obdurate opposition to Israel by Tunisia, whose involvement in Arab-Israeli affairs was always characterized by tolerance, moderation, and sophistication.

Israeli-Mauritanian affairs exist primarily on a symbolic level.   The two countries are a great distance from one another, but Mauritania is of particular interest to Israel because a relationship with Mauritania would add one more Arab state to the growing list of Arab countries with which Israel had concluded an understanding and ultimately diplomatic ties.   Mauritanian-Israeli ties are cordial and constructive given the limitations on these two states.

The above is meant to capture the nature of relations between Israel and the Arab dialogue partners to the South. The key point is that NATO's Mediterranean Initiative was unveiled in the wake of the Arab-Israeli peace effort and thus was premised on a reasonable expectation that tolerable ties between Israel and these states would continue. Unfortunately, the deterioration of Arab-Israeli relations in the wake of Prime Minister Netanyahu's election has had a negative effect not only on the Arab-Israeli peace process but on NATO's Mediterranean Initiative as well.

In the early days following Netanyahu's election, the Likud Party's apparent opposition to the peace process as designed by Shimon Peres seriously undermined ties between Israel and an array of Arab states, including Morocco, Tunisia, Jordan, Egypt, Oman, Qatar, and others. Although the early days of the post-Peres period were particularly damaging, it is important to put this within a broader perspective. Thus, although the heads of many Arab states have little practical sympathy for Yasser Arafat, the bonds of collective Arab identity were sufficiently strong that policies by Israel (or more specifically by the new Netanyahu government) that were perceived to be anti-Palestinian were readily interpreted as being Anti-Arab as well. The Netanyahu government's outspoken attachment to the occupied territories on the West Bank, as well as its willingness to ignore agreements signed by its predecessor, quickly cast a pall over Arab-Israeli ties throughout the region. Israel's unwillingness to withdraw from Hebron exacerbated these tensions even more. At the same time, certain states in Europe, especially France, became involved in Arab-Israeli diplomacy in a way that further kept Arabs and Israelis at a distance from one another. Netanyahu's opposition to the peace process led to a freezing of ties, if not their actual diminution, between Israel and a wide array of Arab states. During this period, there was a highly publicized dispute between Jordan's King Hussein, a key supporter of the peace process, and Prime Minister Netanyahu during a joint visit to the Oval Office in Washington. Morocco and Tunisia froze relations with Israel and refused to upgrade ties with Israel as they had agreed. Oman refused to open a long-promised interest section in Israel and there were serious questions about the future of Israeli-Arab ties in general.

The pendulum briefly swung back when a Hebron agreement was hammered out. Segments of a relieved Arab world immediately began to reward Israel for its willingness to recognize the agreement and many of the frozen ties or arrangements between Israel and several Arab states began to be reinvigorated. This renewed spirit of Arab-Israeli optimism and collaboration was exceedingly short-lived, as a result of the Netanyahu government's decision to initiate a building effort in the Har Homa section of Jerusalem. This led to renewed Arab-Israeli conflict that continues to place the peace process on hold.

The key point here is that Arab-Israeli relations have an important effect on any NATO initiative that places Israel alongside some Arab states and expects them to cooperate with Israel in an arrangement that explicitly excludes other Arab states. This is not meant to suggest that NATO's initiative is doomed to failure, but rather that it is vulnerable to the broader parameters of Arab-Israeli relations, which are at the moment exceedingly volatile and fragile. This indirect linkage to the broader Arab-Israeli peace process is a complicating factor that cannot be ignored and limits the degree of cooperation that can be achieved within NATO's Mediterranean dialogue.

BILATERALISM VERSUS MULTILATERALISM

Several of the dialogue countries are involved in major bilateral diplomatic and political relationships with important NATO member states, particularly the United States. Thus, the U.S. position toward the dialogue must be considered, because if the dialogue merely duplicates U.S. policy, it will have no impact. However, should the dialogue conflict with U.S. regional goals, the United States is likely to oppose it.

Israeli-American relations are critically important to both Jerusalem and Washington, and neither of them is likely to allow any NATO-sponsored activity to interfere with these relations. This is a particularly delicate moment in the history of this relationship because of the ongoing negotiations about the resolution of the Palestine question. And given that the European view of Arab-Israeli issues has at times differed significantly from the position of the United States, it is quite clear that a collective Euro-American agreement on any Middle Eastern issues is unlikely to occur, and thus the dialogue is unlikely to go far on these issues. The same is true for U.S.-Egyptian, U.S.-Jordanian, and U.S.-Moroccan relations. Each of these, in its own way, is important to the United States, and NATO cannot hope to accomplish a great deal without more active U.S. support than is currently the case.

France also has aspirations to play a regional role in the Middle East. France has traditionally maintained strong ties to Morocco and Tunisia, in particular. It is thus sensitive to efforts by NATO to involve itself in the Mediterranean—an area where France has strong historical ties and important national interests. In addition, French-American perspectives on Middle East–related issues often differ, especially lately. Although many of these issues are not likely to be directly broached within the framework of the NATO initiative, they may indirectly affect the prospects for the initiative's success.

CONCLUSIONS

Although the success of NATO's Mediterranean Initiative is far from guaranteed, its failure is also not preordained. In order for it to succeed, attempts must be made to reconcile a history of distrust between the Middle East and the West. This can be accomplished if NATO rigorously defines what it wants the dialogue to accomplish and then invests adequate resources to educate opinion-makers in the dialogue countries about its goals. It must also confront the challenge of close interaction with states that are politically, culturally, historically, and economically quite different from NATO's core membership. The goals of the dialogue must serve the needs of *both* sides. Thus, close attention must be paid to what the Middle Eastern and North African states want from the dialogue, as well as to NATO's ability to marshal adequate resources to make this an appealing policy option for the governments of these states. NATO must also take pains to make clear that the dialogue is sponsored by the organization in its entirety and not just certain members of it. Attempts to coordinate the initiative with the policies of the EU are also desirable as is the ability to strike a balance between bilateral and multilateral approaches. Finally, it is important to acknowledge that there are forces and factors that NATO may not be able to control. The continued deterioration of the Arab-Israeli peace process or the growth of anti-immigration attitudes and xenophobia in parts of Europe will make the dialogue much more difficult to sustain.

Chapter Five

WHITHER NATO'S MEDITERRANEAN INITIATIVE?
CONCLUSIONS AND POLICY RECOMMENDATIONS

Since 1989, NATO has concentrated most of its energy on enlargement to Eastern Europe and internal adaptation; the Mediterranean has received only sporadic attention. Given NATO's other pressing priorities, this benign neglect was understandable. There was a strong need to respond to security problems in Eastern Europe precipitated by the end of the Cold War and the disintegration of the Soviet Union and to adapt the NATO command structure to the radically new security environment. The Bosnia conflict also presented a major security challenge that demanded NATO's attention.

However, in the coming decades, the Mediterranean region is likely to become more important—not just for the Southern members of NATO but for the Alliance as a whole. If NATO and EU enlargement succeed, East Central Europe will be increasingly stable and integrated into Euro-Atlantic political, economic, and security institutions. The real security problems will be on the Alliance's Southern periphery—in the Balkans, the Mediterranean, and the Caucasus.

With the end of the Cold War, the locus of risks and challenges is moving south. At the same time, the distinction between European and Mediterranean security is becoming increasingly blurred as a result of the spillover of economic and social problems from the South, such as immigration, terrorism, and drug trafficking, to Europe. Europe is also more exposed to risks from the Middle East.

The growing turmoil in the South will increasingly affect Alliance interests. In addition, the growing involvement of the EU through the

Barcelona process will have an indirect impact on NATO. As the EU becomes more deeply involved in the Mediterranean region, Mediterranean issues will increasingly become part of the European security agenda—and invariably part of NATO's agenda as well. This will make close coordination between the EU and NATO in the Mediterranean more necessary.

The proliferation of WMD will also thrust Mediterranean issues more forcefully onto the NATO agenda. As noted earlier, within the next decade, all the capitals of Southern Europe could be in range of ballistic missiles launched from North Africa and the Middle East. This will create new security dilemmas for these states—and for the Alliance—and could give the security dialogue with these countries quite a different character.

In addition, a number of the countries along the Mediterranean littoral—particularly Jordan, Egypt, Syria, and Morocco—face political successions. These leadership changes could jeopardize the internal stability of these countries—and the regional stability in the Mediterranean more broadly. At the same time, some countries, such as Algeria and Egypt, face major threats from radical Islamic movements. Hence, the prospect of growing instability and even terrorism being exported to Europe can by no means be excluded.

Finally, the Greek-Turkish dispute over the Aegean and Cyprus is likely to remain a source of concern and keep the Alliance's attention focused on the Mediterranean. As long as these issues remain unresolved, there is always a danger that some incident could unintentionally lead to a new confrontation, as almost happened as a result of the flare-up over the Aegean islet of Imia/Kardak in February 1996.

Thus, for a whole host of reasons, NATO will be forced to pay greater attention to challenges in the Mediterranean. *The real issue, therefore, is not whether NATO should have a Mediterranean policy but what the nature and content of that policy should be and how it can be most effectively implemented.*

NATO'S MEDITERRANEAN INITIATIVE AT THE CROSSROADS

NATO's Mediterranean Initiative, launched at the end of 1994, signaled the Alliance's recognition of the growing importance of the security challenges in the Mediterranean. However, as the discussion in Chapter Three underscores, progress in developing the initiative has been slow.

The low profile adopted by NATO toward the Mediterranean Initiative has been due to a number of factors. First, there is little strong enthusiasm for the initiative within the Alliance, except among the Southern members. Most NATO members have been willing to support the initiative as long as it is limited to "dialogue" and does not require any increased expenditure of resources.

Second, the goals of the initiative remain ill-defined. It is unclear whether the main purpose of the initiative is simply to conduct a dialogue with the countries of the Southern Mediterranean or whether the initiative should also be part of a broader effort to establish defense cooperation with these countries. This lack of clarity, in large part, reflects the limited consensus within NATO about what the initiative is really supposed to do.

Third, NATO has had other pressing priorities, particularly enlargement, internal adaptation, and developing a viable partnership with Russia. Many countries do not want to see attention and scarce resources diverted from enlargement and the PfP.

Fourth, NATO suffers from a serious "image" problem in the dialogue countries, especially among the broader public. The publics in many of these states view NATO as a Cold War institution that is now searching for a new enemy. As a result, the governments in many dialogue countries are wary of cooperating too closely with NATO, especially in the security and defense area, fearing that this will spark a hostile reaction among key segments of their publics.

Fifth, although the dialogue countries have generally responded positively to NATO's Mediterranean Initiative and have indicated their willingness to continue the dialogue with NATO, many are unsure of the purpose of the initiative and do not yet see its relevance to their security problems, most of which are socioeconomic in nature.

Others find the initiative useful but would like to see it go beyond dialogue and develop concrete cooperation in specific areas. Few, however, have thought very systematically about what the cooperation should specifically entail. They expect NATO to take the lead in refining and further elaborating the initiative rather than seeking actively to shape it themselves.

Sixth, the initiative is divorced from NATO's broader security and defense agenda in the Mediterranean. This agenda involves such important security issues as counterproliferation, counterterrorism, peacekeeping, and humanitarian assistance. However, the relationship between this broader security and defense agenda and the Mediterranean Initiative is unclear. Moreover, many of the dialogue countries prefer to concentrate on "soft security" and economic issues rather than on hard security issues. This imposes serious limits on the dialogue with these countries.

Seventh, the deterioration of the Arab-Israeli peace process has diminished the willingness of many Arab countries to engage in a dialogue with Israel. This has inhibited progress in NATO's Mediterranean Initiative and made it more difficult to engage many of the North African and Middle Eastern countries in a dialogue in which Israel is included.

Finally, NATO's Mediterranean Initiative is divorced from the broader U.S. strategic agenda in the Mediterranean and the South more generally. In addition, the United States remains concerned that the initiative could interfere with the Middle East peace process and divert attention from Eastern enlargement. Hence, Washington has expressed only perfunctory interest in the initiative. However, without strong U.S. support, the initiative is unlikely to amount to much.

As a result of all these factors, NATO's Mediterranean Initiative has not really gotten off the ground. It remains largely an afterthought rather than a serious Alliance initiative with strong political support and momentum. If it is to succeed, the initiative needs to be reinvigorated and given stronger political backing, particularly by the United States. It also needs to be more closely harmonized with other initiatives in the region, particularly the Barcelona process.

Otherwise, it is likely to become yet another in a long line of failed Western initiatives in the Mediterranean.

HARD SECURITY VERSUS SOFT SECURITY

In addition, the Alliance is confronted with a major contradiction in its dealings with the countries of the Southern Mediterranean. NATO's comparative advantage is in the area of hard security. The dialogue countries, however, are primarily interested in soft security issues. This raises an important dilemma regarding how to structure the dialogue. What should the main priority in the dialogue be: hard security or soft security?

Given NATO's poor image in many of the dialogue countries and the sensitivity about hard security issues in these states, it may be better in the early stages for NATO to concentrate on soft security and building confidence rather than moving directly to defense and military cooperation with many of the dialogue countries. This is all the more true because many of the member states of NATO, particularly the United States and members in the Southern region, already have established bilateral defense cooperation with the dialogue countries.

The main reason for initially concentrating on soft security is political/psychological. As noted, many of the dialogue countries have strong reservations about NATO. There is thus a need to develop a "bottom-up" approach—to first develop trust and confidence. This can lay the groundwork for the development of concrete military cooperation later on.

The best means for helping to develop such trust and confidence would be to expand considerably the participation of the dialogue countries in seminars on security issues of mutual interest and invite representatives of dialogue countries to attend NATO-sponsored events, including military exercises. This would help to develop greater *transparency* and could lead to the development of a more positive image of NATO in the dialogue countries over time.

This, in turn, could lay the groundwork for the development of military cooperation at a later stage. Such military cooperation, however, should be carried out on *a case-by-case basis*. In effect, the same principle of "self-differentiation" that guides PfP should be used as a guideline for the Mediterranean Initiative. This would recognize the differences between the different dialogue countries and allow military cooperation to develop at its own natural pace. Some countries, such as Egypt and Israel, may be willing to develop some low-level forms of military cooperation, particularly in the areas of civil emergency planning, peacekeeping, and peace support activities, while others may not feel comfortable with such cooperation for quite a while. And some may never want it.

NATO, however, should not be the *demandeur*. Rather, the Alliance should allow cooperation to develop at a pace with which each dialogue country feels comfortable. At the same time, NATO should be careful to ensure that military cooperation with Israel does not get too far out in front of military cooperation with other Arab states, especially Egypt and Jordan. This could undermine the effectiveness and objectives of the initiative.

ARMS CONTROL AND CONFIDENCE-BUILDING MEASURES

Some low-level arms control and CBMs may also contribute to enhancing transparency and building trust. However, the Alliance should recognize that the "security culture" in the dialogue countries differs markedly from that in Europe and that the CSCE/OSCE experience, which has contributed to enhancing security and cooperation in Europe, cannot automatically be transferred to the Middle East and North Africa, where the political and psychological environment is quite different. Many Arab states feel that CSCE/OSCE-type CBMs are premature at present and can be introduced only after the conclusion of a Middle East peace settlement. Otherwise they fear such measures will perpetuate the status quo that the Arabs are trying to change. Moreover, many of the security problems in the Mediterranean are of an internal nature. They are thus not amenable to resolution through classical arms control and confidence-building measures.

This does not mean that CBMs have no utility or should not be tried, but rather that different types of CBMs are needed—at least in the initial stages—from those developed in the CSCE/OSCE context. Rather than trying to introduce arms limitations zones and other similar military CBMs, it might be better to concentrate initially on measures designed to increase transparency and defuse threat perceptions. Such measures include security seminars, educational visits, and inviting dialogue representatives to observe military exercises. This could lay the groundwork for more robust military CBMs—including participation in military exercises—later on.

PfP AND THE MEDITERRANEAN

Some Alliance members have suggested that PfP should be extended to the Mediterranean. They point to the success of PfP in Eastern Europe and argue that extending it to the Mediterranean could help reduce threat perceptions and foster closer cooperation in the region. On the surface, this idea has considerable attraction. PfP has worked remarkably well in Eastern Europe and has been more successful than most observers and NATO officials initially anticipated. Why not extend it to the Mediterranean?

However, in developing cooperation with the countries of the Southern Mediterranean, the Alliance needs to be sensitive to the very different political and cultural environment that exists in North Africa and the Middle East in comparison with that in Eastern Europe. What works in Eastern Europe may not necessarily work or be desirable in the Middle East and North Africa. In Eastern Europe, NATO has a very positive image. It is regarded as the main pillar of the new security order in Europe and as a guarantee against any residual security threat that might be posed at some point by a resurgent Russia. It is also a symbol of membership in the Western "club," which the countries of Eastern Europe all want to join. Indeed, many East European countries see PfP as a stepping stone to NATO membership.

The situation in the dialogue countries is quite different. These countries endured years of colonialism and they remain deeply suspicious of the West. Many see NATO as an instrument of Western military intervention and domination. They fear that with the disappearance of the former Soviet Union, NATO is looking for a new enemy and that Islam could become that new enemy. While these perceptions may be inaccurate, they are deeply held in many quarters in the dialogue countries and they strongly color the attitude of elites toward cooperation with NATO.

In addition, there is not a uniform and commonly accepted defini-tion of "security" on the two sides of the Mediterranean. When rep-resentatives of the Maghreb and Middle East think of security, most think first and foremost of internal rather than external security. This lack of an agreed "language" and common approach to security hinders the establishment of a security dialogue with the countries of North Africa and the Middle East and often contributes to misper-ceptions and misunderstandings on both sides.

In short, PfP cannot be transferred lock, stock, and barrel to the Mediterranean. While some aspects of PfP may be applicable to the Mediterranean, in general, the political diversity of the region will re-quire separate and specific solutions appropriate to the region.

THE IMPACT OF THE MIDDLE EAST PEACE PROCESS

NATO's ability to move forward with its Mediterranean Initiative will also be affected by external factors, particularly the Middle East peace process. While there is no direct formal linkage between NA-TO's Mediterranean Initiative and the peace process, developments within the Middle East peace process affect the willingness of dia-logue countries to cooperate with Israel and even participate in multilateral meetings where Israel is present. They also affect the tone and tenor of the meetings themselves. Thus, the two dialogues cannot be entirely separated even if no formal linkage between the two exists.

The deterioration of Arab-Israeli relations since the election of Prime Minister Benjamin Netanyahu in May 1996, for instance, has blocked progress in the Barcelona process, particularly in the security area.[1] It has also affected the MENA summits—the Doha summit in Qatar was boycotted by most of the Arab countries, including Egypt and Morocco. As long as the Middle East peace process remains stalled, it will be difficult to intensify cooperation with the dialogue countries in many areas, especially on sensitive security issues.

---

[1]The Euro-Med summit in Malta in April 1997 was largely overshadowed by Arab-Israeli polemics. For details, see "Middle East Overshadows Malta Talks," *Financial Times*, April 16, 1997. See also Judy Dempsey, "Euro-Mediterranean Forum Stalls," *Financial Times*, December 15, 1997.

MEMBERSHIP IN THE DIALOGUE

At present, six countries are included in NATO's Mediterranean Initiative: Egypt, Israel, Morocco, Tunisia, Jordan, and Mauritania. However, there is pressure both from the dialogue countries as well as from some NATO members to include other countries along the Southern Mediterranean littoral.

In principle, there is no fundamental obstacle or objection to expanding the dialogue. The NATO Council's approach has been that the initiative is open and can be progressively expanded on a case-by-case basis. Jordan, for instance, joined at the end of 1995 and, in principle, the door is open to others as well.

However, the experience of the EU's Euro-Mediterranean Partnership (the Barcelona process) suggests that there are merits in keeping the dialogue small and focused. Once the dialogue becomes too large, it is hard to achieve a consensus among participants, especially on sensitive defense and security issues. Moreover, the larger the membership, the more likely it is that extraneous issues, such as the Arab-Israeli dispute, will be injected into the dialogue, making progress difficult.

Some NATO members, particularly Spain, have suggested that Algeria should be invited to join the initiative. They argue that it is hard to have a serious dialogue about Mediterranean security without including Algeria, which is the largest country in the Maghreb and one of the most important exporters of gas and energy to Europe, especially to Southern Europe. In addition, Algeria is a member of the WEU Mediterranean dialogue and the EU Barcelona process.

However, there are risks to inviting Algeria to join the dialogue at this stage. Many members of the Algerian government see such an invitation as a means of confirming the legitimacy of the regime—something NATO may not necessarily want to do at the moment. Thus, until the political situation in Algeria is more settled, it would be better to postpone any consideration of inviting Algeria to join the dialogue.

Syria could be a potential candidate down the road. However, including Syria in the dialogue before an Israeli-Syria peace accord would risk bringing the Arab-Israeli conflict more directly into the dialogue and could paralyze it. The same is true for Lebanon, which is essentially a Syrian client. The problems encountered within the Euro-Mediterranean Partnership dialogue on arms control, confidence-building, and conflict prevention should serve as a warning in this regard.[2]

Some dialogue countries would also like to see Libya included. However, the United States and a number of other NATO members are strongly opposed to including Libya because of its support for international terrorism. Moreover, Libya is not a participant in the EU or WEU Mediterranean dialogues. Until Libya moderates its international behavior, especially its support for terrorism, it should not be considered for membership.

BILATERALISM VERSUS MULTILATERALISM

To date the dialogue has been largely bilateral, except in the field of information. Many of the dialogue countries do not seem ready for multilateral discussions, especially as long as relations with Israel are strained. Moving too quickly to multilateralize the dialogue, therefore, could be disruptive and reinforce existing tensions within the group. Thus, for the moment, it would seem advisable to continue most activities on a bilateral basis. However, some activities could be conducted on a multilateral basis. Multilateral seminars devoted to security issues, for instance, can contribute to creating a broader security community in the Mediterranean and fostering greater regional cooperation in a variety of areas. Some MCG meetings could also take place at the multilateral level.

COORDINATION WITH OTHER INITIATIVES

NATO also needs to consider how the initiative fits into the broader pattern of other ongoing efforts aimed at enhancing cooperation and security in the Mediterranean, discussed in Chapter Two. NATO's

---

[2]On these problems, see in particular Roberto Aliboni, Abdel Monem Said Aly, and Alvaro de Vasconcelos, *Joint Report of the EuroMeSCo Working Groups on Political and Security Cooperation and Arms Control, Confidence-Building and Conflict Prevention*, April 1997.

efforts should be designed to complement and reinforce these other efforts, not duplicate them.

Given the fact that the main security problems in the region are internal and have their roots in economic, social, and political factors, it makes sense for the EU to take the lead in dealing with the Mediterranean. The EU is much better equipped than NATO to deal with these problems. NATO should not try to duplicate the EU's efforts. Rather it should try to focus on those areas in the security/military field where it has a comparative advantage.

At the same time, the large number of institutions and organizations dealing with Mediterranean security issues increases the importance of broadly coordinating NATO's efforts with those of other institutions, especially of the EU and WEU, to avoid duplication. The political and military dialogues conducted by NATO and the WEU, for instance, have largely the same goals and include nearly the same countries.[3] Security issues, especially CBMs, are also part of the Euro-Mediterranean dialogue within the Barcelona process.

The EU has traditionally jealously guarded its independence and has been reluctant to coordinate its efforts with NATO. Thus, initially there may be considerable bureaucratic resistance within the EU—as well as NATO—to coordinating the two dialogues. But greater consultation and coordination between the two institutions is essential. Otherwise there is a danger that the two dialogues could undercut, rather than complement, one another.

This coordination need not require the establishment of an elaborate institutional mechanism. It could be achieved by an informal periodic exchange of views—perhaps two or three times a year—between the Secretary General of NATO and the head of the EU Commission and/or Commissioners dealing with Mediterranean issues. Such an informal periodic exchange of views would allow high-level officials in each institution to keep the others abreast of what each institution is doing in the Mediterranean and to ensure that the two dialogues are broadly complementary.

---

[3]The WEU dialogue includes Algeria but not Jordan. Otherwise the memberships are the same.

AREAS FOR EXPANDED COOPERATION

In addition, a number of steps in specific areas could be taken to give the Mediterranean Initiative new momentum and relevance. Most of these steps would not require a major expenditure of additional resources. Four areas in particular should be given top priority:

- Public information and outreach

- Educational courses and visits

- Civil emergency planning (CEP)

- Crisis management, peacekeeping, and peace support.

In particular, NATO's public information and outreach effort needs to be expanded. This is critical if NATO is to change perceptions in the dialogue countries and create better understanding of the Alliance's goals and purposes. It is also an essential building block for broader cooperation in other areas over the long run. At the same time, NATO should do more to involve dialogue countries in civil emergency planning, educational courses and visits, and crisis management, peacekeeping, and peace support activities. These are areas where NATO has a comparative advantage over other institutions and where many of the dialogue countries have expressed an interest in greater cooperation.

The discussion below focuses on specific steps NATO could take in these and other areas that contribute to stability in the Mediterranean over the long run.

Political Cooperation

Use the Initiative to Increase Understanding of NATO's Goals and Purposes. NATO has a poor image in many of the dialogue countries. NATO, therefore, should use the Mediterranean Initiative to give the political elites and opinion-makers in the dialogue countries a better sense of NATO's true goals and purposes. Special emphasis should be put on the way NATO is adapting to a new security environment and NATO's new missions, especially peace support activities. This is particularly important because many elites in the dialogue countries fear that these activities are designed to create a new intervention force in the Mediterranean aimed against them.

Tie the Dialogue More Closely to NATO's Agenda. NATO should use the dialogue with the Mediterranean states to further its own goals in the Mediterranean. Issues such as counterproliferation, terrorism, and drug trafficking should become an integral part of NATO's dialogue with the countries of the Middle East and North Africa. This can help to raise the consciousness of the dialogue countries about these problems and promote discussion of ways to more effectively address these problems over the long run.

Make Counterproliferation a Key Element of the Political Dialogue. In the coming decade, the proliferation issue will take on increased importance in the Mediterranean, as both dialogue countries and Southern European members of NATO become more vulnerable to weapons of mass destruction. NATO needs to begin to engage the countries of the Southern Mediterranean in a dialogue about how to deal with this emerging threat. At the same time, NATO needs to be careful that the dialogue does not become a vehicle for an all-out assault on Israel over the Nonproliferation Treaty.

Develop Concrete Forms of Cooperation and Work Plans. Currently NATO holds bilateral discussions with representatives of the dialogue countries twice a year, although each side can request additional meetings if it so desires. So far, these meetings have largely been limited to an exchange of information about NATO's goals and activities. After two years, these discussions, while useful, have pretty much run their course. Many countries are now looking to more concrete forms of cooperation. The dialogue therefore needs to be go beyond mere get-to-know-you sessions and exchanges of information and begin to develop more concrete forms of cooperation. NATO could, for instance, begin to develop concrete work plans, modeled after those developed for the East European and Commonwealth of Independent States (CIS) countries within the framework of the NACC. This would help to give the initiative greater focus and concrete content.

Adapt Work Plans to Needs of Individual Countries. Some dialogue countries will be more ready to proceed with concrete cooperation than others. As with PfP, the principle of "self-differentiation" should be adopted. This would allow different countries to proceed at *their own pace*. However, in contrast to PfP, with the dialogue countries there would be no attempt to restructure military forces or prepare the dialogue countries for NATO membership. In some cases, in fact, there might be no concrete defense cooperation or military-related activities at all. In others there might be some military-related activities, especially in the area of peacekeeping. In other words, the pace and content of the dialogue and military-related activities would primarily depend on the desires of the individual dialogue country.

Scientific Cooperation

Invite and Financially Support the Participation of Representatives of Dialogue Countries in NATO-Sponsored Scientific Meetings. Currently, dialogue countries are being invited to scientific meetings on a self-funding basis. However, given the economic constraints facing these countries, few representatives of dialogue countries have the means to attend NATO scientific meetings unless their attendance is subsidized. Therefore, NATO would need to allocate funds to underwrite the participation of representatives from the dialogue countries at these meetings.

Sponsor Several Meetings Per Year on Scientific Subjects of Particular Interest to the Dialogue Countries. Some of these meetings could be held in dialogue countries. Such meetings could act as a kind of CBM and would demonstrate NATO's multifaceted dimensions as well as help to disseminate scientific knowledge of value and use to the dialogue countries.

Information

Expand Information Activities to Dialogue Countries. Over the last three years, the NATO Office of Information and Press has focused its information efforts on opinion-makers and research and defense institutes in the dialogue countries. In 1997, NATO's Office of Information and Press supported two major conferences, one international seminar and one visit to NATO by key opinion-makers from the dialogue countries. However, if the objectives of NATO's initiative are to be achieved, the information outreach to the dialogue countries needs to be significantly expanded and more closely coordinated with other activities of NATO member states in the Mediterranean.

The officer in charge of Information activities for the Mediterranean dialogue countries in the NATO Office of Information and Press should be given increased resources to enable him to interact more frequently and intensively with opinion-makers from the dialogue countries as well as with his EU and WEU counterparts dealing with the Mediterranean. Given the contacts that he has already established with opinion-makers in the dialogue countries over the last three years, this officer should also act as the coordinator of NATO's broader information effort toward the Mediterranean countries.

Increase Support for Visits to NATO by Key Opinion-Makers from the Dialogue Countries, Especially Journalists, Academics, and Parliamentarians. These leaders play a major role in shaping public opinion in their respective countries. At the same time, more systematic effort needs to be made to identify the key opinion-makers in the dialogue countries and to acquaint them with NATO's purpose and goals.

Focus Attention on Emerging, Younger Elites. NATO should focus particular attention on emerging, younger elites. These new elites will move into leadership positions in the coming decade. Moreover, many of them are more open and less hostile to NATO than the previous generation that grew up during the colonialist period or the Cold War.

Translate and Disseminate Key NATO Materials and Documents in Arabic. NATO could publish an Arabic version of *NATO Review*. It should also consider producing a documentary film on NATO's purposes, goals, and transformation *in Arabic*. Production and dissemination of such material in Arabic would underscore NATO's seriousness about reaching out to the Arab world. These materials would reach a much broader audience and complement NATO's other efforts aimed at increasing contacts with opinion-makers, the media, and research institutes in the dialogue countries. The cost of producing and disseminating these materials would be relatively modest, while the potential long-term benefits in terms of shaping public opinion in the Arab world could be substantial.

Develop and Strengthen Ties to Research and Defense Institutions in the Dialogue Countries. NATO should continue developing ties to major defense and research institutes in the dialogue countries. The main goal of these ties should be to help develop a strong and vital security community in the dialogue countries and increase the knowledge and understanding of defense and security issues in these countries. At present, such a security community barely exists in most dialogue countries, with the exception of Israel (and to a lesser extent, Egypt). This inhibits both the pace and the tenor of the dialogue with the states of the Maghreb and Mashreq. In many cases, the two sides are speaking an entirely different "language" and have different definitions of security. Developing closer ties and promoting exchanges with institutes in the dialogue countries could help to narrow this gap and facilitate the establishment of a more productive security environment over the long run. To this end, the NATO Office of Information and Press should be allowed to conduct joint projects with these institutes in the dialogue countries themselves. The NATO Defense College should also be encouraged to establish closer contacts with these institutes.

Sponsor Fellowships and Exchanges for Researchers from Dialogue Countries at Major Institutes in NATO Countries Dealing with Defense and Security Matters. A certain number of NATO fellowships each year could also be allocated to scholars from the dialogue countries and used to support research dealing with Mediterranean security issues. For instance, a special fellowship program similar to the Manfred Wörner Fellowship program for the PfP countries could be set up for the dialogue countries of the Mediterranean.

Researchers and scholars who receive such awards should be encouraged to focus their research on such topics as regional arms control and CBMs in the Mediterranean, proliferation problems, the effects of arms sales on regional security, and the establishment of regional security regimes in the Mediterranean. The fellowship and exchange program should be designed not only to encourage an exchange of ideas between scholars in the dialogue countries and those in NATO member states, but also to create a vital and viable "security community" in the dialogue countries over the long run.

Pay Greater Attention to Mediterranean Security Issues in NATO's Own Publications. A special issue of the *NATO Review*, for instance, could be devoted to Mediterranean security. Leading scholars and officials from the dialogue countries could be invited to contribute to the issue. The issue could also be published *in Arabic*. This would ensure a wider audience in the Arab world and would also underscore NATO's seriousness about its commitment to the dialogue.

Develop a Coordinated Public Information Effort for STANAV-FORMED Port Visits. Port visits by STANAVFORMED can play a useful role in developing contacts in the dialogue countries and acquainting the general publics in these countries with NATO's goals and purposes. However, if these visits are to be successful, they need to be accompanied by a coordinated public information campaign. Otherwise there is a danger that the visits may be misperceived or even arouse opposition.

Problems in carrying out public information activities associated with the port visits have occurred in the past because the dialogue countries do not have direct diplomatic relations with NATO.[4] In some cases, poor coordination between the Foreign Ministry and Defense Ministry in the dialogue countries has created problems and impaired the success of the visits. While efforts have recently been made to reduce these problems, greater attention to coordinating public information aspects of the visits in advance would enable NATO to exploit the visits to influence public attitudes in the dialogue countries. Also, greater effort needs to be made to coordinate STANAVFORMED port visits with other NATO-related as well as bilateral activities in the dialogue countries.

To ensure coordination of these various information efforts, the officer in charge of information activities for the Mediterranean countries in the NATO Office of Information and Press should be put in charge of NATO's broader information effort. He or she should act as

---

[4]During a visit by STANAVFORMED to Morocco in September 1996, for instance, the AFSOUTH Public Information Officer (PIO) tried to carry out a public information effort to support the visit. The PIO was not able to talk to the Moroccan authorities directly, but had to go through the Embassy of a NATO country in Rabat because the Moroccan Ministry of Defense asserted that Morocco did not have diplomatic relations with NATO. While these problems were eventually overcome, ship visits by the public were not possible since the ships were moved inside a controlled area open only to visitors in advance. Other STANAVFORMED visits have been confronted with similar problems in this regard, making it difficult to carry out an effective public information campaign to support the visits.

the main coordinator and central clearing house for all NATO-related information activities. He or she should also be tasked with tracking bilateral activities of NATO member states with the dialogue countries (exercises, port visits, military exchanges, etc.)—activities that affect NATO's relations with the dialogue countries.

Visits/Courses

Encourage Greater Participation in the Courses, Especially Peacekeeping Courses, at the NATO School in Oberammergau. At present, representatives from the dialogue countries can attend the courses at the NATO School in Oberammergau on a self-funding basis. Few dialogue countries, however, can afford to send participants on this basis. If NATO is serious about its commitment to the initiative, it will have to subsidize the participation of representatives from the dialogue countries for these courses—at least in the beginning.

Open Up the Academic Programs at the NATO Defense College in Rome, Such As the Senior Course and General/Flag Officers' Course, to Representatives of the Dialogue Countries. As with the courses at the NATO School at Oberammergau, the main problem is that few dialogue countries are in a position to pay for such courses. Therefore, NATO should—at least in the initial stages—underwrite the costs of their attendance. Moreover, the curriculum should be worked out jointly between representations of the dialogue countries and representatives of NATO Defense College. This would give the countries a greater sense of participation as well as help to ensure that the curriculum included topics of interest to the dialogue states.

Civil Emergency Planning (CEP)

Increase Participation of the Dialogue Countries in CEP Activities Related to Disaster Management As Well As in Courses at the NATO School in Oberammergau. Civil emergency planning provides a fruitful avenue for cooperation. Many of the dialogue countries have expressed continued interest in cooperation in CEP. Moreover, it is an area where NATO has a comparative advantage over other institutions dealing with the Mediterranean. NATO should exploit this comparative advantage by involving the dialogue countries in CEP-related activities and in courses at the NATO school in Oberammergau. Furthermore, cooperation in CEP can serve as a confidence-building measure and lay the groundwork for cooperation in other areas. However, given the financial constraints faced by the dialogue countries, it will be extremely difficult for representatives from the dialogue countries to participate in these activities and courses on a self-funding basis. NATO should therefore be prepared to offer financial assistance to enable them to participate more actively in these activities and seminars.

Crisis Management, Peacekeeping, and Peace Support Activities

NATO's real comparative advantage lies in the military area. It is here that the Alliance can offer expertise that no other institution can. However, given the political sensitivities and perceptions of NATO in many of the dialogue countries, NATO needs to be careful how it approaches the issue of military cooperation. Moreover, these sensitivities are greater in some dialogue countries than others. What is possible in one country in the military area may not be possible in another. Thus, NATO needs to approach the question of military cooperation on a case-by-case and country-by-country basis.

This being said, the following are a number of things that NATO can do.

Expand Cooperation in Peacekeeping. Three dialogue countries—Egypt, Jordan, and Morocco—participated in IFOR/SFOR. NATO should build on this experience. NATO could expand the participation of dialogue countries in peacekeeping courses at the NATO school in Oberammergau and the NATO Defense College in Rome. Dialogue countries could also be invited to send observers to NATO peacekeeping exercises and training camps, including those in prospective new member countries, such as Poland and Hungary.

Invite Dialogue Countries to Send Observers to Large-Scale NATO Exercises. There is currently a great deal of misunderstanding in the dialogue countries about NATO's goals and purposes. One way to correct these misperceptions would be to invite the dialogue countries to send observers to NATO exercises. This goal would increase transparency and understanding of what NATO does—*and does not* *do*—in the military field. This would not only enhance the level of understanding of NATO activities among the dialogue countries but would also help to reduce threat perceptions and make clear that NATO is not planning for an intervention against a "new enemy" in the Middle East.

Invite Officers from the Dialogue Countries to Special Briefings at SHAPE Headquarters in Mons. In addition, officers from the dialogue countries could be invited to special briefings at SHAPE Headquarters in Mons. This would contribute to greater transparency and understanding of NATO's goals and purposes. Eventually officers from dialogue countries might be permanently attached to SHAPE if the cooperation develops sufficiently and if there is sufficient interest on the part of individual dialogue countries.

Adopt a Regional Approach to Military Cooperation. Not all activities need to be done on a NATO-wide basis. NATO simply provides an umbrella for cooperation. Much can be done at the regional level without involving NATO as a whole. Thus, NATO members who have the interest and are motivated to do so can put together a "regional coalition" and undertake military activities that include non-NATO members. They would not need "approval" from NATO to do this. They can take the initiative on their own. For instance, several Southern members of NATO—say Spain, Italy, and Morocco—could conduct military or peacekeeping activities together. Such activities—and the number of countries involved—could gradually be expanded as the dialogue countries got used to working together with NATO members.

This regional approach could provide a fruitful avenue for cooperation with the dialogue countries. It would allow Southern members of NATO—and other members who were interested—to take the initiative and invite select dialogue countries to participate in particular military and/or peacekeeping activities on a regional basis. At the same time, it would allow for a degree of self-differentiation. Not all dialogue countries would have to—or want to—participate in such activities. But some dialogue countries might. For instance, each year Egypt and the Arab Emirates participate in joint exercises (Bright Star) with the United States, France, and the United Kingdom.

This approach would provide a flexible means of involving some dialogue countries in military and peacekeeping activities on a regional basis. As the countries got used to working together, this cooperation could be expanded to other areas. Other dialogue countries or NATO members might decide to join, or new regional coalitions could be formed for other activities.

This approach would also make it easier for the dialogue countries to legitimize and justify such cooperation with NATO to their publics. Dialogue countries would not be conducting exercises or peacekeeping activities with "NATO" but with regional neighbors (who also happen to be members of NATO). Thus, the regional approach would provide a certain degree of political "cover" and help to desensitize cooperation. At the same time, it would help to build regional trust and cooperation and serve as a building block for broader cooperation at a later date.

Finally, it would allow cooperation to develop at its own "natural" pace. No dialogue country would be under any pressure to participate in particular activities if it did not want to. Some would want to participate; others, for a variety of reasons, might not. Cooperation could thus be developed gradually, without discrimination or pressure, on a case-by-case basis.

Arms Control/Verification

Arms control and arms control verification also provide other possible areas for cooperation with the dialogue countries. Many of them lack expertise in the arms control field. NATO could facilitate exchanges of information by taking the following steps.

Invite the Dialogue Countries to Participate in General Arms Control Information Courses at the NATO School in Oberammergau. This would help to build expertise in the dialogue countries and also facilitate a dialogue on regional arms control issues.

Use the NATO Fellowship Program to Support Research on Arms Control in the Mediterranean Region, Especially Research by Specialists from the Dialogue Countries Themselves. The goal would be to expand the pool of knowledgeable specialists on arms control issues as part of the broader effort to create a security community in the Mediterranean.

Establish a Joint Working Group on CBMs. This group could be composed of representatives from the dialogue countries and select NATO members and would be tasked with examining and formulating CBMs in the Mediterranean region. It could draw on, but should not duplicate, the work being carried out on CBMs within ACRS and the Barcelona process. The latter dialogue has not progressed very far, particularly because of the strong opposition from Syria and Lebanon. A NATO-sponsored effort would involve a smaller, less diverse group of countries and would not involve Syria and Lebanon— the two main obstacles to progress on CBMs in the Barcelona process. Thus it might have a better chance of success than the Barcelona effort. Moreover, NATO has more expertise in this area than the EU. It is therefore in a better position to carry out such a dialogue. However, to avoid duplication, NATO should loosely coordinate its CBM efforts with those of the EU.

THE ROLE OF THE MEDITERRANEAN COOPERATION GROUP

The newly created Mediterranean Cooperation Group (MCG), established at the Madrid Summit in July 1997, should be the main body for developing and coordinating NATO's Mediterranean Initiative (except in the field of public information, which, as suggested earlier, should be coordinated by the officer in charge of Mediterranean affairs in the NATO Office of Information and Press). The MCG should be tasked to undertake a major review of NATO's Mediterranean Initiative and to recommend areas where cooperation and dialogue can be further developed. In addition, the MCG should meet periodically—perhaps twice a year—with the dialogue countries to review progress and develop new initiatives.

RESOURCES AND FUNDING

One of the obstacles to an expansion of the Mediterranean Initiative to date has been the unwillingness of NATO members to commit major resources to the Mediterranean Initiative at a time when greater outlays are needed for other important activities, especially enlargement and military operations in Bosnia.

However, if the Mediterranean Initiative is to succeed, NATO will have to devote greater financial resources to it. Even the modest efforts suggested in this study will require some increase in funding, especially in the information area. It is unrealistic, for instance, to expect the dialogue countries to send representatives to courses at the NATO School in Oberammergau and the NATO Defense College in Rome on a self-funding basis. Given the economic constraints these countries face, their defense establishments will be far more likely to spend scarce resources on bilateral exchanges, which are often reciprocal and where the benefits are more tangible. Thus, NATO will have to find ways to subsidize participation of the dialogue countries in these courses and to expand its outreach, especially in the information field.

TOWARD A BROADER SOUTHERN STRATEGY

The steps recommended above essentially represent incremental adjustments to what NATO is already doing rather than a radical reorientation of the Alliance's policy in the Mediterranean. However, if the Mediterranean Initiative is going have a significant impact over the long run, it must be linked to a broader Alliance strategy toward the South. Otherwise the initiative will remain largely an afterthought, disconnected from the Alliance's broader agenda in the South, and it will be difficult to obtain the support of key Alliance members outside the Southern region.

This strategy will need to link a number of diverse elements into a coherent whole:

- command reform
- PfP (for candidate and noncandidate members)
- policy toward the Balkans
- dialogue with the countries along the Southern Mediterranean littoral

- enhancing cooperation and stability in the Aegean
- counterproliferation.

At present, these various elements are running on different tracks and are not closely connected to one another. They need to be integrated into a broader, more coherent strategy designed to enable NATO to better meet the security challenges it is likely to face in the coming decades.

In particular, the role of AFSOUTH needs to be upgraded as part of the overall effort to modernize and streamline NATO's command structure. During the Cold War AFSOUTH was of secondary importance. The Central Front was where the action was. Hence it made sense that AFCENT should get the lion's share of NATO resources.

With the end of the Cold War, however, the responsibilities and importance of AFSOUTH have increased significantly. During the Cold War, the AFSOUTH commander (CINCSOUTH) had to worry primarily about the Soviet Mediterranean fleet. Today his Area Of Strategic Interest (AOSI) includes the Balkans, the Mediterranean, parts of the Gulf, and the Caucasus. Yet there has been little corresponding shift in resources to enable the AFSOUTH commander to carry out his expanded responsibilities or to closely monitor and plan for contingencies in his AOSI.

This imbalance needs to be redressed as NATO carries out the process of internal adaptation. More resources need to be shifted to AFSOUTH to enable the AFSOUTH commander to carry out his expanded responsibilities and to monitor and plan for contingencies in his AOSI. In addition, the command reform under way needs to be designed to meet the new challenges NATO is likely to face in the future, many of which are likely to be in the South.

At the same time, the Alliance needs to ensure that its political and military strategy in the Mediterranean are closely harmonized. Efforts to enhance NATO's power projection capabilities could create new anxieties and fears among the dialogue countries and inhibit efforts to intensify cooperation with them. Thus, it is important that any changes in NATO's military strategy and command structure be carefully explained to these countries ahead of time to reduce the chances of misperception and misunderstanding.[5]

Embedding the Mediterranean Initiative in a broader Southern strategy would also help to ensure stronger U.S. support. To date, the United States has not exhibited a strong interest in the initiative, in part because it does not see the linkage between the initiative and many of the "big" strategic issues in the South. The more NATO's initiative can be linked to this broader U.S. agenda in the Mediterranean, the more likely it is to obtain backing in Washington. Such support will be crucial if the initiative is to really gain political weight and momentum.[6]

---

[5]The strong reaction of some dialogue countries to the opening of the EUROFOR headquarters in Florence in November 1996 illustrates the problems in this regard and underscores the need to ensure that NATO's political and military strategy are closely harmonized.

[6]The highly visible U.S. role in promoting NATO enlargement to the East highlights this point. Without strong U.S. backing and active engagement, it is unlikely that enlargement would have become a major NATO priority.

SELECT BIBLIOGRAPHY

Aliboni, Roberto, "Institutionalizing Mediterranean Relations: Complementarity and Competition," *Internationale Politik und Gesellschaft*, Nr. 3, 1995, pp. 90–99.

——, "Instability South of the Mediterranean," *The International Spectator*, July–September 1993, pp. 87–96.

Aliboni, Roberto, Abdel Monem Said Aly, and Alvaro de Vasconcelos, *Joint Report of the EuroMeSCo Working Groups on Political and Security Cooperation and Arms Control, Confidence-Building and Conflict Prevention*, April 1997.

Asmus, Ronald, F. Stephen Larrabee and Ian O. Lesser, "Mediterranean Security: New Challenges, New Tasks," *NATO Review*, May 1996, pp. 25–29.

Bensidoun, Isabelle and Agnes Chevallier, *Europe-Méditerranée: Le Pari de L'Ouverture*, Economica, Paris, 1996.

Bos, Eduard, et al., *World Population Projections 1994–95*, Baltimore, Md.: Johns Hopkins University Press, 1994.

Boyer, Yves, et al., "Europe and the Challenge of Proliferation," *Chaillot Paper No. 24*, Western European Union (WEU) Institute for Security Studies, Paris, May 1996.

Braudel, Fernand, *The Mediterranean and the Mediterranean World in the Age of Philip II*, New York: Harper and Row, 1966.

Calleya, Stephen C., *Navigating Regional Dynamics in the Post–Cold War World*, Dartmouth, Brookfield, Vermont, 1997.

Callies de Salies, Bruno, "Méditerranée: Quelle Politique Envers les Etats du Sud?," *Défense Nationale*, February 1996, pp. 93–107.

Colard, Daniel, "La Conférence de Barcelone et le Partenariat Euro-Méditerranéen," *Défense Nationale*, February 1996, pp. 109–117.

de Rato, Rodrigo, *In Pursuit of Euro-Mediterranean Security*, Working Group on the Southern Region, North Atlantic Assembly, May 1996.

de Vasconcelos, Alvaro, *Européens et Maghrébins*, Editions KARTHALA, Paris, 1993.

Drauzan, Jean-François, and Raoul Girardet, *La Méditerranée, Nouveaux défis, nouveaux risques*, Éditions PUBLISUD, Paris, 1995.

European Commission, "The Barcelona Conference and the Euro-Mediterranean Association Agreements," European Commission Information Memorandum, 23 November 1995.

———, "The European Union's Mediterranean Policy," European Commission Memorandum 94/64, October 1994.

———, "The European Union's Relations with the Mediterranean," European Commission Memorandum 94/74, December 1994.

European Union, "Strengthening the Mediterranean Policy of the European Union," *Bulletin of the European Union*, Supplement, 2/95,p. 17.

Faria, Fernanda, and Alvaro de Vasconcelos, "Security in North Africa: Ambiguity and Reality," *Chaillot Paper No. 25*, WEU Institute for Security Studies, September 1996.

Feldman, Shai, *Nuclear Weapons and Arms Control in the Middle East*, Cambridge, Mass.: MIT Press, 1997.

Fuller, Graham E., *Algeria: The Next Fundamentalist State?* Santa Monica, Calif.: RAND, MR-733-A, 1996.

Fuller, Graham E., and Ian O. Lesser, *A Sense of Siege: The Geopolitics of Islam and the West*, Boulder, Colo.: Westview, 1995.

Hadar, Leon T., "Meddling in the Middle East?" *Mediterranean Quarterly*, Fall 1996.

Halliday, Fred, *Islam and the Myth of Confrontation: Religion and Politics in the Middle East*, London: I. B. Tauris, 1996.

Hoekman, Bernard and Simeon Djankov, "The European Union's Mediterranean Free Trade Initiative," *The World Economy*, July 1996, pp. 387–406.

Holmes, John W., ed., Maelstrom: *The United States, Southern Europe and the Challenges of the Mediterranean*, Cambridge, Mass.: World Peace Foundation, 1995.

Huntington, Samuel, "The Clash of Civilizations," *Foreign Affairs*, Summer 1993, pp. 22–49.

Ibrahim, Saad Eddin, "Competing Visions of the Arab Middle East," (unpublished), December 1996.

Jean, Carlo, "Security in the Mediterranean and Italy's Role," *Mediterranean Quarterly*, Spring 1997, pp. 129–145.

Jentleson, Bruce, *The Middle East Arms Control and Regional Security (ACRS) Talks: Progress, Problems, and Prospects*, San Diego, Calif.: Institute on Global Conflict and Cooperation (IGCC) Policy Paper No. 26, September 1996.

Joffe, George, "Integration of Peripheral Dependence: The Dilemma Facing the Southern Mediterranean States," paper presented to the Conference on Cooperation and Security in the Mediterranean After Barcelona, Mediterranean Academy of Diplomatic Studies, Malta, 22–23 March 1996.

Kaplan, Robert D., "The Coming Anarchy," *The Atlantic Monthly*, February 1994, pp. 44–76.

———, *The Ends of the Earth: A Journey at the Dawn of the 21st Century*, New York: Random House, 1996.

Khallaf, Hani, "Enlargement of Western Security Institutions: A Non-Western Perception," (unpublished).

Kruzel, Joseph, "The New Mediterranean Security Environment," presentation at the Institute of National Strategic Studies (INSS)/AFSOUTH Conference on Mediterranean Security, Naples (mimeograph), February 27–March 1, 1995.

Lesser, Ian O., *Mediterranean Security: New Perspectives and Implications for U.S. Policy*, Santa Monica, Calif.: RAND, R-4178-AF, 1992.

———, *Security in North Africa: Internal and External Challenges*, Santa Monica, Calif.: RAND, MR-203-AF, 1993.

Lesser, Ian O., and Ashley J. Tellis, *Strategic Exposure: Proliferation Around the Mediterranean*, Santa Monica, Calif.: RAND, MR-742-A, 1996.

Lorca, Alejandro V., and Jesus A. Nuñez, "EC-Maghreb Relations: A Global Policy for Center-Periphery Interdependence," *The International Spectator*, July–September 1993, pp. 53–66.

Moya, Pedro, *Cooperation for Security in the Mediterranean: NATO and EU Contributions*, Subcommittee on the Mediterranean Basin, North Atlantic Assembly, May 1996.

———, *Frameworks for Cooperation in the Mediterranean Basin*, Subcommittee on the Mediterranean Basin, North Atlantic Assembly, October 1995.

———, *NATO's Role in the Mediterranean*, Subcommittee on the Mediterranean Basin, North Atlantic Assembly, April 16, 1997.

Mureddu, Guiseppe, *Approvvigionamento delle Materie Prime e Crisi e Conflitti nel Mediterraneo*, Rivista Militare, Rome, 1993.

Pierre, Andrew J., and William B. Quandt, *The Algerian Crisis: Policy Options for the West*, Washington, D.C.: Carnegie Endowment Books, 1996.

Rhein, Eberhard, "Mit Geduld und Ausdauer zum Erfolg," *Internationale Politik*, 2/96, pp. 15–20.

Serfaty, Simon, "Algeria Unhinged: What Next? Who Cares? Who Leads?" *Survival*, Vol. 38, No. 4, Winter 1996–97, pp. 137–153.

Solana, Javier, "NATO and the Mediterranean," *Mediterranean Quarterly*, Spring, 1997, pp. 11–20.

Steinberg, Gerald M., "European Security and the Middle East Peace Process," *Mediterranean Quarterly*, Winter 1996, pp. 65–80.

———, "Disintegration and Integration in the Mediterranean," *The International Spectator*, July–September 1993, pp. 67–78.